MATERIALS SCIENCE FOR SOLAR ENERGY CONVERSION SYSTEMS

RENEWABLE ENERGY SERIES

Editor-in-Chief: A A M Sayigh

Pergamon Titles of Related Interest

BANHIDI
Radiant Heating Systems: Design and Applications

BEI
Modern Power Station Practice, 3rd edition

HARRISON
Geothermal Heating

HORIGOME
Clean and Safe Energy Forever

McVEIGH
Sun Power, 2nd edition

SAITO
Heat Pumps

SAYIGH
Energy Conservation in Buildings

SAYIGH
Energy and the Environment: into the 1990s, 5-vol set

SAYIGH & McVEIGH
Solar Air Conditioning and Refrigeration

STECCO & MORAN
A Future for Energy

TREBLE
Generating Electricity from the Sun

Pergamon Related Journals (*free specimen copy gladly sent on request*)

Energy
Energy Conservation and Management
Geothermics
Heat Recovery Systems and CHP
International Journal of Heat and Mass Transfer
International Journal of Hydrogen Energy
Progress in Energy and Combustion Science
Renewable Energy
Solar Energy

MATERIALS SCIENCE FOR SOLAR ENERGY CONVERSION SYSTEMS

Edited by
C. G. GRANQVIST
Physics Department,
Chalmers University of Technology, Sweden

PERGAMON PRESS
OXFORD · NEW YORK · SEOUL · TOKYO

U.K.	Pergamon Press plc, Headington Hill Hall, Oxford OX3 0BW, England
U.S.A.	Pergamon Press Inc., 395 Saw Mill River Road, Elmsford, NY 10523, U.S.A.
KOREA	Pergamon Press Korea, KPO Box 315, Seoul 110-603, Korea
JAPAN	Pergamon Press, 8th Floor, Matsuoka Central Building, 1-7-1 Nishi-Shinjuku, Shinjuku-ku, Tokyo 160, Japan

---

First edition 1991

Library of Congress Cataloging in Publication Data

Materials science for solar energy conversion systems/edited by C. G. Granqvist.—1st ed.
p. cm.—(Renewable energy series)
1. Solar energy—Materials. I. Granqvist, Claes G. II. Series.
TJ812.7.M4 1991 621.47—dc20 91-811

British Library Cataloguing in Publication Data
Granqvist, C. G.
Materials science for solar energy conversion systems. -
(Renewable energy)
I. Title II. Series
621.47028

ISBN 0-08-040937-7

Printed in Great Britain by BPCC Wheatons Ltd, Exeter

CONTENTS

FOREWORD

Rapid advances in materials science now make possible new vistas in solar energy conversion systems. Thus today's technology allows high quality surface coatings to be used in energy-efficient fenestration and in spectrally selective solar absorbers. Bulk materials are available for transparent thermal insulation and for fluorescent solar concentrators. Chromogenic windows, polymeric light-pipes, and radiative cooling materials are examples of technologies that hold promise for rapid future development. Chemical materials and photochemical processes are being seriously considered for energy storage.

This book concentrates on new advances in materials science with particular attention to applications, performance, characterization, laboratory manufacturing, industrial production, reliability, and cost. The book consists of an introductory chapter and five topical chapters dealing with different, yet highly integrated, subjects in materials science as applied to solar energy conversion systems. The first topical chapter is *Optical Properties of Inhomogeneous Two-Component Materials* by G.A. Niklasson. This theoretical treatise provides a background to much of the basic work on spectrally selective radiative properties in the subsequent chapters. *Transparent Insulation Materials* by Platzer and Wittwer deals with different approaches to convection suppression in transparent materials. These novel materials have a high potential for increasing the efficiency of solar thermal conversion systems and of other energy-related applications. *Selectively Solar-Absorbing Surface Coatings: Optical Properties and Degradation* by G.A. Niklasson and C.G. Granqvist introduces selective absorption of solar energy and discusses the properties of currently available surface coatings, with particular consideration of recent progress in the understanding of degradation phenomena. *Energy-Efficient Windows: Present and Forthcoming Technology* by C.G. Granqvist presents up-to-date research and development that will provide important opportunities for improved energy-efficient windows. This chapter also covers the design criteria for different climates and reviews means to fulfil these through proper materials selection. Concepts such as large-area chromogenics and angular selectivity - that have been forwarded only during the past few years - are discussed. The last chapter, *Materials for Radiative Cooling to Low Temperatures* by C.G. Granqvist and T.S. Eriksson, introduces the idea of employing the clear sky as a heat sink for radiative cooling. It is shown that under certain conditions a temperature difference can be obtained using suitably designed materials: some results of selected field tests are included.

Each of the contributors has set a landmark in the field of materials science, particularly with regard to renewable energy utilization. We hope that this first book in the 'Renewable Energy' series, *Materials Science for Solar Energy Conversion Systems,* will be invaluable to all scientists, engineers, and industrialists working in this and related areas.

A.A.M. Sayigh

Chapter 1

INTRODUCTION TO MATERIALS SCIENCE FOR SOLAR ENERGY CONVERSION SYSTEMS

C.G. Granqvist

Physics Department
Chalmers University of Technology and University of Gothenburg
S-412 96 Gothenburg, Sweden

ABSTRACT

Several pathways for renewable energy conversion are introduced. Materials for specific solar energy applications have optical properties tailored to the requirements set by the radiation in our surroundings. This "natural" radiation is outlined, and the goals of materials science for several solar energy conversion systems are discussed.

I. RENEWABLE ENERGY CONVERSION PATHWAYS

The limited availability of fossil and nuclear fuels, and their environmental impacts, have led to a growing awareness of the importance of renewable energy sources. Political considerations and incidental market fluctuations may have short term effects, but they will not offset the tendency that renewable energy sources, and the materials for implementing their associated benign technologies, are going to play an ever-increasing role both in the industrialized and less developed countries. Given this situation, materials science for renewable energy conversion systems - which this book is all about - is sure to be of growing importance. The topic lies at the crossroads of basic physics and chemistry, materials fabrication, and energy technology. Besides providing great intellectual challenges to the materials scientist, this field offers personal satisfaction: its goals are not to develop means of mass destruction or luxury items for the affluent few, but rather to promote sustainable development and a decent quality of life for all humankind.

The importance of renewable energy sources are currently attracting widespread attention. One manifestation of this interest is the recent report by the World Commission on Environment and Development (also known as the "Brundtland Report"), which states that[1] *renewable sources /.../ should form the foundation of the global energy structure during the 21st century. Most of these sources are currently problematic but given innovative development, they could supply the same amount of primary energy the planet now consumes. However, achieving these use levels will require a programme of coordinated research, development, and demonstration projects /.../ to ensure the rapid development of renewable energy.*

The present book can be viewed as an attempt to give some materials science input into the research and development called for in the Brundtland Report.

Figure 1 gives a schematic representation[2] of renewable energy conversion. Solar energy can be converted into useful forms through pathways in the geosphere, biosphere and technosphere. Thus mass flow in the atmosphere can be used for windpower, and water flow can be used for hydroelectric/hydromechanic power, wave power, and tidal power. Heat gradients in the sea may be useful for ocean thermal energy conversion (OTEC). In the biosphere, solar energy is required for photosynthesis leading to food production and to energy conversion based on biomass and biogas. The technosphere gives a multitude of options for manmade collectors of solar energy and for energy-efficient passive design in architecture.[3] Among the collectors, one can distinguish between those utilizing thermal conversion ("solar collectors") and quantum conversion ("solar cells").

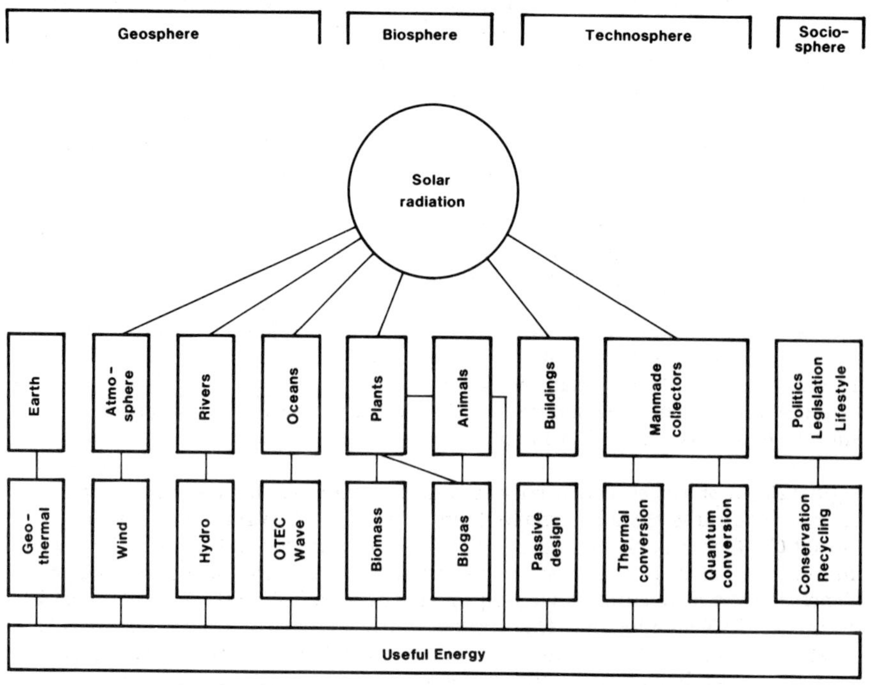

Fig. 1. Renewable energy conversion pathways. (From Ref. 2).

Geothermal energy - which does not rely on solar radiation - is a viable source of energy in certain locations. Finally, political decisions and legislative actions, as well as individual conviction, can lead to a "non-consumptionist" lifestyle involving energy conservation and materials recycling and hence to a decrease in the demand on primary energy.

From Fig. 1 it should be obvious that very many kinds of materials can be employed for renewable energy conversion. Thus, for example, new high strength polymeric and metallic materials are of interest for wind energy and for OTEC installations. It follows that it is not practical to consider materials for all of the applications mentioned in Fig. 1 as a self-contained group. However, one can single out "materials for solar energy conversion" as a category whose properties are tailored specifically to meet the requirements set by the spectral content and intensity of the radiation in our natural surroundings.[3] Such materials, many of which are most useful as thin surface coatings, are discussed in this book. More specifically, the book mainly considers the science of materials of relevance for thermal conversion in manmade collectors and for energy-efficient passive design in architecture.

II. NATURAL RADIATION

The basic principles of efficient solar energy conversion systems can be grasped only if one has a clear idea of the radiation that prevails in our natural surroundings. This radiation is introduced in Fig. 2, where the different spectra are drawn with a common logarithmic wavelength scale.[3]

All matter emits radiation. The properties of this radiation are conveniently discussed by starting with the ideal blackbody, whose emitted spectrum - known as the Planck spectrum - is uniquely defined if the absolute temperature is known. Planck's law is a consequence of the quantum nature of radiation. Part (a) of Fig. 2 depicts Planck spectra for four temperatures. The vertical scale denotes power per unit area and wavelength increment (hence the unit GW m^{-3}). The spectra are bell-shaped and confined to the $2 < \lambda < 100$ μm range. The peaks in the spectra are displaced toward shorter wavelength as the temperature goes up; this is referred to as Wien's displacement law. At room temperature the peak lies at about 10 μm. Thermal radiation from a material is obtained by multiplying the Planck spectrum by a numerical factor - the emittance - which is less than unity. In general, the emittance is wavelength dependent.

Figure 2(b) reproduces a solar spectrum for radiation outside the earth's atmosphere.[4] The curve has a bell shape defined by the sun's surface temperature (~6000°C). One observes that the solar spectrum is limited to the $0.25 < \lambda < 3$ μm interval, so that there is almost no overlap with the spectra for thermal radiation. Hence one can have surfaces whose properties are entirely different with regard to thermal and solar radiation. The integrated area under the curve gives the solar constant (1353 ± 21 W m^{-2}); this is the largest possible power density on a surface oriented perpendicular to the sun in the absence of atmospheric extinction.

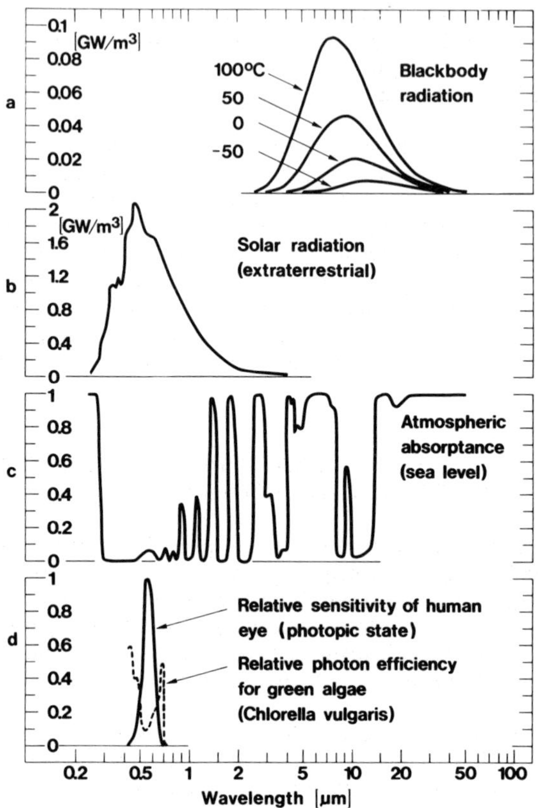

Fig. 2. Spectra for (a) blackbody radiation pertaining to four temperatures, (b) solar radiation outside the earth's atmosphere, (c) typical absorptance across the full atmospheric envelope, (d) relative sensitivity of the human eye and relative photon efficiency of green algae. (From Ref. 3).

The solar energy conversion systems of present concern are to be placed at ground level, and it is of obvious interest to consider to what extent atmospheric absorption influences solar irradiation and thermal emission. Figure 2(c) illustrates a typical absorption spectrum vertically across the full atmospheric envelope at clear weather conditions.[5] The spectrum is found to be quite complicated with bands of high absorption - caused mainly by water vapour, carbon dioxide, and ozone - and intervening bands of high transparency. It is evident that the majority of the solar radiation can be transmitted down to ground level, and only parts of the ultraviolet ($\lambda < 0.4$ μm) and infrared ($\lambda > 0.7$ μm) tails are strongly damped. The maximum power density perpendicular to the sun is limited to about 1000 W m^{-2}. Thermal radiation from a surface exposed to the clear sky is seen to be strongly absorbed except in the $8 < \lambda < 13$ μm range, where the transmittance can be large provided that the humidity is moderately low. The thermal radiation can be large in the 8-13 μm interval, and hence one concludes that a non-negligible part of the emitted energy can go straight through the atmosphere. This phenomenon constitutes the basis for radiative cooling.

Figure 2(d) illustrates two biophysical constraints of interest for applications. The solid curve shows the relative spectral sensitivity of the human eye in its light-adapted (photopic) state. The bell-shaped curve extends across the $0.4 < \lambda < 0.7$ μm interval with its peak at 0.555 μm.[6] In its darkness-adapted (scotopic) state, the eye's sensitivity is displaced about 0.05 μm toward shorter wavelengths. Hence a large part of the solar energy comes as infrared radiation. Photosynthesis in plants operates with wavelengths in approximately the same range as those for the human eye, which is of obvious relevance for greenhouse applications. An example of the relative photon efficiency for green algae is given by the dashed curve.

III. SOME GOALS OF MATERIALS SCIENCE FOR SOLAR ENERGY CONVERSION SYSTEMS, AND BOOK OUTLINE

The different types of ambient radiation are spectrally selective, i.e., confined to well-defined and often non-overlapping wavelength ranges, as apparent from Fig. 2. This is of major significance and, in fact, the goal of materials science for solar energy conversion systems is to develop materials - often surface coatings - which take advantage of the specific features of the ambient radiation. The following properties and materials are of particular relevance:

(1) High transmittance of solar radiation can be achieved in materials with porous or cellular configurations that prevent air circulation. Such materials suppress convective heat transfer and promote thermal insulation.

(2) High absorption of solar radiation can be combined with low emittance of thermal radiation by use of spectrally selective surfaces. These materials are of interest for efficient photothermal conversion of solar energy.

(3) High transmittance of solar radiation can be combined with low emittance of thermal radiation, and high transmittance of visible light can be combined with high reflectance of infrared solar radiation. Surface coatings with these properties are of great interest for energy-efficient fenestration in cold and hot climates, respectively.

(4) Materials with highly angular-dependent radiative properties can be used in several different ways in energy-efficient fenestration.

(5) Coatings with radiative properties that can be changed to match differing demands of heating and lighting are of obvious interest for achieving energy-efficient buildings. The relevant materials have "chromogenic" properties.[7]

(6) The high atmospheric transmittance in the 8-13 μm wavelength range can be used for channelling energy from sky-facing surfaces into space. Thus it is possible to devise materials for efficient passive cooling to temperatures well below the air temperature.

Items (1) - (6) are covered in several topical chapters below. They are preceded by a theoretical treatise on *Optical Properties of Inhomogeneous Two-Component Materials* by G.A. Niklasson. This chapter discusses matters of interest for understanding some basic materials science discussed in the topical chapters. However, it is not the case that a detailed understanding of the - sometimes rather involved and technical - exposition in the theory chapter is a prerequisite for grasping the topical presentations. The first topical chapter is related to item (1); it is entitled *Transparent Insulation Materials* and is authored by W. Platzer and V. Wittwer. Item (2) is treated in a discussion on *Selectively Solar-Absorbing Surface Coatings: Optical Properties and Degradation* by G.A. Niklasson and C.G. Granqvist. Items (3)-(5) are given a unified and extensive presentation entitled *Energy-Efficient Windows: Present and Forthcoming Technology* by C.G. Granqvist. The final chapter of the book is related to item (6) and covers *Materials for Radiative Cooling to Low Temperatures;* it is authored by C.G. Granqvist and T.S. Eriksson.

The aim of the book is to give a detailed and up-to-date presentation of several key aspects of materials science for solar energy conversion systems, excluding those relying on quantum conversion. It should be remembered, though, that even with this restriction there are several interesting and important materials that are not discussed; among these are reflector materials, light concentrators, thermal storage media, and others.

REFERENCES

1. World Commission on Environment and Development, *Our Common Future* (Oxford University Press, Oxford, UK, 1987), p. 15.
2. C.G. Granqvist, in *Energy and the Environment into the 1990s*, edited by A.A.M. Sayigh (Pergamon, Oxford, UK, 1990), Vol. 3, p. 1465; Appl. Phys. A 52, 83 (1991).
3. C.G. Granqvist, *Spectrally Selective Surfaces for Heating and Cooling Applications* (SPIE Opt. Engr. Press, Bellingham, USA, 1989).
4. M.P. Thekaekara, in *Solar Energy Engineering*, edited by A.A.M. Sayigh (Academic, New York, 1977), p. 37.
5. K. Ya. Kondratyev, *Radiation in the Atmosphere* (Academic, New York, 1969).
6. G. Wyszecki and W.S. Stiles, in *Color Science*, 2nd edition (Wiley, New York, 1982), p. 256.
7. C.M. Lampert and C.G. Granqvist, editors, *Large-area Chromogenics: Materials and Devices for Transmittance Control* (SPIE Opt. Engr. Press, Bellingham, USA, 1990); C.G. Granqvist, Crit. Rev. Solid State Mater. Sci. 16, 291 (1990).

Chapter 2

OPTICAL PROPERTIES OF INHOMOGENEOUS TWO-COMPONENT MATERIALS

G.A. Niklasson

Physics Department
Chalmers University of Technology and University of Gothenburg
S-412 96 Gothenburg, Sweden

ABSTRACT

Effective medium and multiple scattering theories for the optical properties of two-component materials are reviewed. Such materials have numerous applications in the field of coatings for energy efficiency. The transmittance and reflectance of a coating or slab of a composite can be obtained from the effective dielectric and magnetic permeabilities of the material. For materials with inhomogeneities much smaller than the wavelength of the impinging radiation, the effective dielectric permeability can be evaluated in the quasistatic limit. We review the rigorous Bergman-Milton bounds for the effective dielectric permeability as well as various effective medium theories that have been put forward for describing the optical properties of specific microstructures. Specifically we treat the effects of pair and three-point correlations on the bounds and obtain novel effective medium theories taking these effects into account. Materials with large inhomogeneities on the order of, or larger than, the wavelength must be described by different theories. The effective magnetic permeability must be taken into account. The specular reflectance, the direct transmittance and the diffuse scattering are treated by use of a four flux theory.

I. INTRODUCTION

An understanding of the optical properties of inhomogeneous materials is very important in the development and optimization of various coatings for energy efficiency. Applications such as solar absorption, radiative cooling and energy efficient windows have prompted a large interest in composite materials. Many coatings used for selective absorption of solar energy are of this class.[1] Composites of metal particles in an insulator matrix display a very good selectivity, and have been produced by electrochemical techniques,[2-5] electron-beam evaporation[6,7] and sputtering.[8] Another example of composite selective absorbers are paint coatings[9] which consist of an absorbing pigment dispersed in a binder material.

For radiative cooling applications solar reflecting and infrared-transmitting pigmented polymer foils are of interest, as well as ceramics which probably consist of a mixture of an oxide phase and voids.[10] As a final example we mention that many visibly transmitting coatings with low emittance - of use for energy-efficient windows - incorporate thin metal films. It has recently been shown that an improved performance can be achieved by inhomogeneous "network" films close to the percolation threshold.[11] The aim of this chapter is to review theoretical descriptions of the optical properties of two-component materials. These theories are often very useful in the study of materials for energy-efficient applications. However, this chapter is not a prerequisite for the rest of the book, and the reader mainly interested in technical applications may proceed directly to the topical chapters.

In the development of composite thin films it is of prime importance to establish the connection between the properties of the composite and those of the constituents. This facilitates materials selection and optimization of practical coatings. The optical properties of composite materials can be described in the quasistatic approximation if the size of the inhomogeneities is much smaller than the wavelength of electromagnetic radiation. The optical properties of the material are obtained from the effective dielectric permeability of the composite which can be related to the dielectric permeabilities of the constituents by effective medium theories (EMT's).[12-14] These theories are also sensitively dependent upon the actual microgeometry of the composite material. Actually a rigorous expression for the effective dielectric permeability can only be obtained if the detailed geometrical arrangement (i.e., the n-point correlation functions) of the constituents are known. When limited structural information is available, as is always the case in practice, the various EMT's can give no more than approximate expressions. However it is possible to obtain rigorous bounds for the effective dielectric permeability.[15-18] When more structural information is incorporated into the bounds, they become more narrow. For large size inhomogeneities the quasistatic approximation is not valid and the concept of an effective medium encounters difficulties.[19] Scattering becomes important and the optical properties can be described in the framework of radiative transfer and multiple scattering theory.[20-22] In the case of very large particles, simplifications are again possible and geometric optics can be used.

In this chapter the various theories that have been put forward to describe the optical properties of composite materials are reviewed. We will consider materials with inhomogeneities of any size, but our treatment is restricted to two-component materials. In Sec. II below we make a brief description of thin film optics. Here the transmittance and reflectance of a thin film are related to the dielectric and magnetic permeabilities of the film material. Some models of the dielectric permeability, which give insight into the physical phenomena involved, are also described. In Sec. III we treat the case of small inclusions of the components of the composite material. The effective dielectric permeability is then evaluated in the quasistatic approximation. We describe the rigorous bounds and effective medium theories that are valid in this limit, and consider cases where different amounts of information about the composite is known, namely the dielectric permeabilities and volume fractions of the constituents and isotropy of the structure. This leads to the Wiener,[15] Hashin-Shtrikman[16] and Bergman-

Milton[17,18] bounds. Simple effective medium theories for the dielectric permeability are applicable to special microstructures.

Recent advances in the characterization of composite materials have made it practical to incorporate more information about the geometry than the volume fractions and the condition of isotropy into rigorous bounds and effective medium theories. In Sec. IV we treat the situation when the pair and three-point correlation functions are known. This kind of theory has so far only been applied to certain cases, i.e., to fractal structures[23] and random mixtures of hard[24-26] or penetrable[27] spheres in a matrix.

In Sec. V we treat the case of inhomogeneities with larger sizes, where the quasi-static approximation is not applicable. One may derive extended effective medium theories that describe some aspects of the optical properties in this case.[19] However, when scattering is of importance a completely different approach is necessary, and radiative transfer or multiple scattering theory[20-22] has to be used. It turns out that in many cases, e.g. when considering coatings, important simplifications can be made. A four-flux theory that is sufficiently accurate for comparisons with spectrophotometric data on inhomogeneous materials is described. Some final remarks are made in Sec. VI.

II. THIN FILM OPTICS AND THE DIELECTRIC PERMEABILITY.

When studying the optical properties of a material one generally measures the reflectance and transmittance as a function of wavelength and angle of incidence. These quantities are functions of the dielectric and magnetic permeabilities of the material. Conversely, it is possible to evaluate for example the dielectric permeability from carefully chosen combinations of experimental transmittance and reflectance data.[28] In this section the theory of the optical properties of a thin film on a substrate is described. As a first step we consider light incident towards the boundary between two media denoted i and j. The angle to the surface normal is θ_i, as indicated in Fig. 1. The media are characterized by their complex dielectric and magnetic permeabilities, $\varepsilon_{i,j}$ and $\mu_{i,j}$. Part of the light is reflected at the boundary(r^{ij}) and part is transmitted (t^{ij}) through the boundary. We distinguish between light with s-polarisation (E vectors normal to the plane spanned by the incident, reflected and transmitted beams) and with p-polarisation (H vectors normal to the same plane). From Maxwell's equations, one can obtain the well known Fresnel's relations for the reflected field amplitudes:[29]

$$r_s^{ij} = \frac{n_i \cos\theta_i - \frac{\mu_i}{\mu_j}(n_j^2 - n_i^2 \sin^2\theta_i)^{1/2}}{n_i \cos\theta_i + \frac{\mu_i}{\mu_j}(n_j^2 - n_i^2 \sin^2\theta_i)^{1/2}}, \tag{1}$$

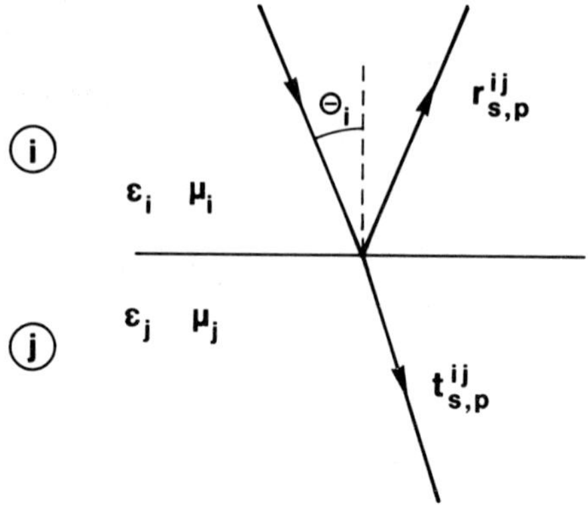

Fig. 1. Definitions of symbols entering Fresnel's relations.

$$r_p^{ij} = \frac{\frac{\mu_i}{\mu_j} n_j^2 \cos\theta_i - n_i (n_j^2 - n_i^2 \sin^2\theta_i)^{1/2}}{\frac{\mu_i}{\mu_j} n_j^2 \cos\theta_i + n_i (n_j^2 - n_i^2 \sin^2\theta_i)^{1/2}} , \tag{2}$$

$$t_s^{ij} = \frac{2n_i \cos\theta_i}{n_i \cos\theta_i + \frac{\mu_i}{\mu_j} (n_j^2 - n_i^2 \sin^2\theta_i)^{1/2}} , \tag{3}$$

$$t_p^{ij} = \frac{2n_i\, n_j \cos\theta_i}{\frac{\mu_i}{\mu_j} n_j^2 \cos\theta_i + n_i (n_j^2 - n_i^2 \sin^2\theta_i)^{1/2}} . \tag{4}$$

Here n_i and n_j denote the refractive indices of the media; they are given by

$$n_i = (\varepsilon_i\, \mu_i)^{1/2} , \tag{5}$$

and analogously for n_j. Fresnel's relations can be used to discuss the optical properties of a thin film on a substrate. We consider the geometry specified in Fig. 2 and let (2) denote the film (of thickness d) and (3) the substrate. A medium (1) surrounds the coated substrate. Further, we let (f) signify light incident from the frontside and (b) signify light incident from the backside. Equations (1) - (4) yield expressions for r^{12}, r^{23}, t^{12}, t^{21}, t^{23}, t^{32}, r^{31} and t^{31}. Neglecting the effect of the backside of the substrate, the amplitude reflectance and transmittance for the film, r_2 and t_2, are obtained from[30]

$$r_{2s}^{f} = \frac{r_s^{12} + r_s^{23}\, e^{2i\delta}}{1 + r_s^{12}\, r_s^{23}\, e^{2i\delta}} , \tag{6}$$

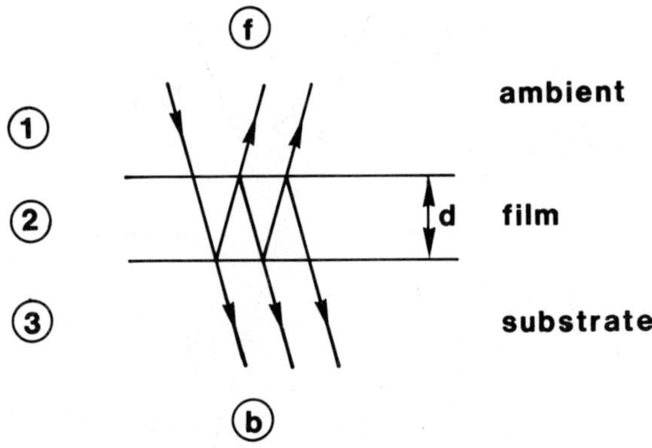

Fig. 2. Geometry used for discussing the optics of a thin film on a substrate.

$$r_{2s}{}^{f} = \frac{t_s{}^{12}\, t_s{}^{23}\, e^{i\delta}}{1 + r_s{}^{12}\, r_s{}^{23}\, e^{2i\delta}} \, . \tag{7}$$

Relations of the same form are easily found for $r_{2s}{}^{b}$, $t_{2s}{}^{b}$, $r_{2p}{}^{f,b}$, $t_{2p}{}^{f,b}$ as well. In Eqs. (6) and (7), δ is the phase change of the light beam upon traversing the film. Specifically[30]

$$\delta = \frac{2\pi d}{\lambda} \left(\varepsilon_2\, \mu_2 - \varepsilon_1\, \mu_1 \sin^2\theta_1\right)^{1/2} \, , \tag{8}$$

where λ denotes the wavelength of light. The measurable optical quantities are the light intensities. They are denoted by capital letters, i.e., for example $R_{2s}{}^{f}$,and are given by the formulae[30]

$$R_{2s,p}{}^{f,b} = |\, r_{2s,p}{}^{f,b}\, |^2 \, , \tag{9}$$

$$T_{2s}{}^{f} = \frac{(\varepsilon_3\, \mu_1)^{1/2} \cos\theta_3}{(\varepsilon_1\, \mu_3)^{1/2} \cos\theta_1} \, |\, t_{2s}{}^{f}\, |^2 \, , \tag{10}$$

$$T_{2s}{}^{f} = \frac{(\varepsilon_1\, \mu_3)^{1/2} \cos\theta_3}{(\varepsilon_3\, \mu_1)^{1/2} \cos\theta_1} \, |\, t_{2s}{}^{f}\, |^2 \, . \tag{11}$$

In addition the expressions for $T_{2s}{}^{b}$ and $T_{2p}{}^{b}$ are obtained from $t_{2s}{}^{b}$ and $t_{2p}{}^{b}$ by interchanging the indices 1 and 3 in Eqs. (10) and (11) above.

If the substrate is metallic, the transmittance is zero and only the reflectance has to be considered. In this case it is immediately given by $R_{2s,p}{}^{f}$. If the substrate is transparent, a more complicated situation exists, and multiple reflections in the substrate must be taken into account. These are incoherent for large substrate thicknesses and must be included through addition of the intensities of the multiply reflected beams. The final expressions for the reflectance and the transmittance are

$$R_s = R_{2s}{}^f + \frac{T_{2s}{}^f\, T_{2s}{}^b\, R_{3s}}{1 - R_{2s}{}^b\, R_{3s}}\,, \tag{12}$$

$$T_s = \frac{T_{2s}{}^f\, T_{3s}}{1 - R_{2s}{}^b\, R_{3s}}\,, \tag{13}$$

where $R_{3s,p}$ and $T_{3s,p}$ are obtained from $r_{s,p}{}^{31}$ and $t_{s,p}{}^{31}$ by relations analogous to Eqs. (9) - (11).

The treatment above can easily be generalized to multiple layer films.[30] This is most conveniently done by use of the characteristic matrix technique .

We have now related the transmittance and reflectance of a thin film on a substrate to the dielectric and magnetic permeabilities. In general, these quantities are complex and wavelength dependent. For homogeneous materials in the ultraviolet, visible and infrared wavelength ranges one may set the magnetic permeability μ equal to unity.[31] This is however not generally the case for longer wavelengths, for example in the microwave region.[31] Magnetic permeabilities different from unity must also be included in certain inhomogeneous materials as we come back to in Sec. V below.

The dielectric permeability often shows a complicated behaviour in the infrared range and at shorter wavelengths. The frequency dependent dielectric permeability is expressed by

$$\varepsilon(\omega) = \varepsilon_1(\omega) + i\varepsilon_2(\omega). \tag{14}$$

We next discuss the general form of $\varepsilon(\omega)$ for different *homogeneous* materials. In many cases one can consider this function to be a sum of individual contributions originating from different elementary excitations. In terms of susceptibilities $\chi^i \equiv \chi_1{}^i + i\chi_2{}^i$, one can write[32]

$$\varepsilon = 1 + \chi^{VE} + \chi^{PH} + \chi^{FC}\,, \tag{15}$$

where VE = valence electrons, PH= phonons, and FC = free carriers (usually electrons). The various susceptibilities can easily be distinguished if their resonances fall in well separated wavelength regions. For a particular χ^i, the contribution far from its resonance is real and constant.

Figure 3 gives a schematic representation of the χ^is. The real and imaginary parts are consistent with the Kramers-Kronig relations. For χ^{VE} and χ^{PH}, one can often represent the susceptibilities by a sum of damped Lorentz oscillators, i.e., by expressions of the kind

$$\chi^{Lorentz} = \frac{\Omega_0^2}{\omega_L{}^2 - \omega^2 - i\omega\Gamma}\ , \tag{16}$$

where Ω_0 is the oscillator strength, ω_L is the resonance frequency and Γ represents the width of the resonance peak. For most insulators and good metals, χ^{VE} is resonant in the ultraviolet or the blue part of the visible spectrum, while χ^{PH} is resonant in the thermal infrared. For χ^{FC} one can make use of the Drude theory, at least for a first-order description. The susceptibility can be written

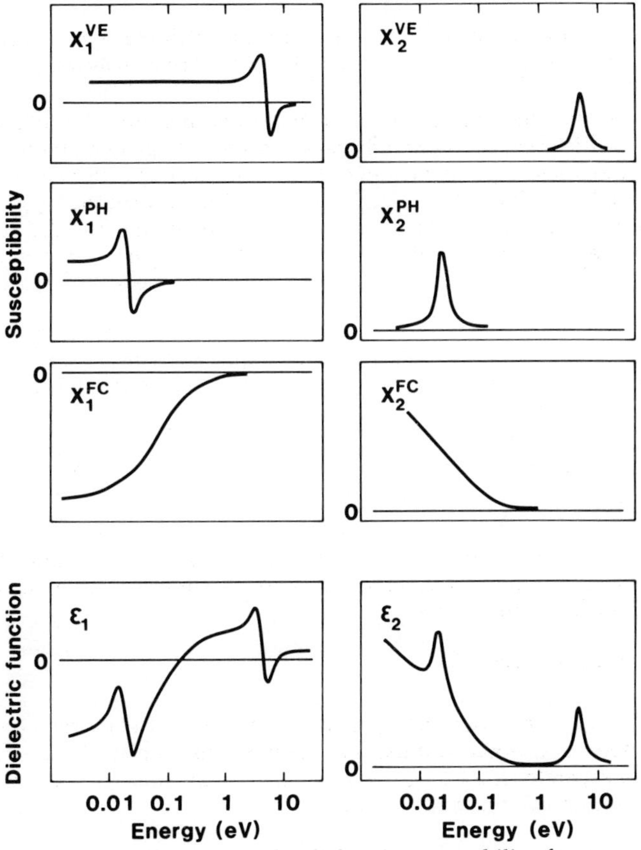

Fig. 3. Contributions to the dielectric permeability from susceptibilities due to valence electrons (VE), phonons (PH) and free carriers (FC).

$$\chi^{Drude} = -\frac{\omega_p^2}{\omega^2 + i\omega\gamma}, \tag{17}$$

where ω_p is the plasma frequency and γ represents the width of the resonance. More elaborate free-electron theories may be required for fully quantitative assessments. It is seen that χ^{Drude} can be obtained by setting $\omega_L = 0$ in the expression for $\chi^{Lorentz}$. Below ω_p, χ_1^{FC} becomes strongly negative while χ_2^{FC} becomes strongly positive, as apparent from Fig. 3. The location of ω_p depends on the free electron density. For a metal, ω_p is normally in the ultraviolet. For a doped semiconductor, ω_p can be in the infrared. In a metal, χ^{PH} is usually not apparent owing to the dominating influence of χ^{FC}. The lowest part of Fig. 3 illustrates the ensuing performance of ε for the case of a heavily doped semiconductor.

The dielectric permeability of many metals, semiconductors and insulators have been derived from optical measurements. Extensive tabulations can be found in different handbooks[33,34], at least for the more common homogeneous materials. In the remainder of this chapter we will be concerned with methods for obtaining the permeabilities of inhomogeneous materials from the permeabilities of the homogeneous constituents.

III. RIGOROUS BOUNDS AND EFFECTIVE MEDIUM THEORIES

In this section we consider the optical properties of inhomogeneous materials in the quasistatic approximation. This is valid when the size of the inhomogeneities or particles is much less than the wavelength of light, or more precisely when $2\pi\varepsilon_m^{1/2} a/\lambda << 1$. Here a is the particle radius and ε_m denotes the largest of the dielectric permeabilities of the materials. In the quasistatic case the optical properties of the material are uniquely described by an effective dielectric permeability, $\bar{\varepsilon}$. Because the field is almost constant over a particle, electrostatic arguments can be used to derive $\bar{\varepsilon}$. We now present rigorous bounds on $\bar{\varepsilon}$ that hold when different amounts of structural information is known, as well as effective medium theories that are fairly good approximations for many practical composite materials. Our discussion is restricted to three-dimensional two-component composites.

We first consider the situation when only the dielectric permeabilities of the two phases, ε_A and ε_B, are known. The bounds on the dielectric permeability in this case are given by the extremal microstructures shown in Fig. 4. They are layered structures oriented in different directions with respect to the applied electric field. The volume fractions of the two phases, f_A and f_B, are defined in Fig. 4. When the field is oriented parallel to the layers the structure is equivalent, in an electrostatic picture, to many capacitors connected in parallel. Hence we obtain for the effective dielectric permeability of this structure

$$\bar{\varepsilon}_{B1} = f_A\,\varepsilon_A + f_B\,\varepsilon_B\ . \tag{18}$$

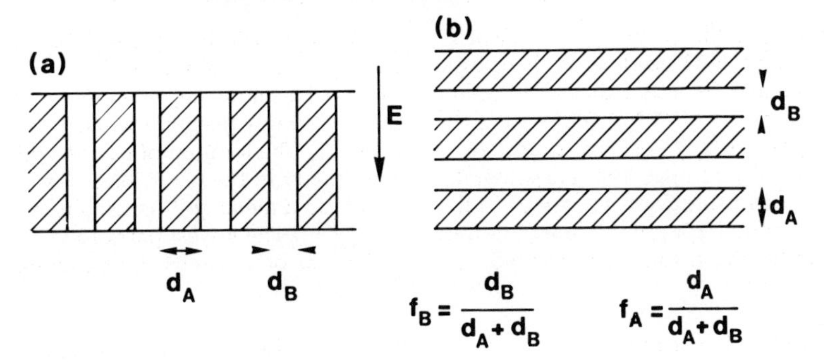

Fig. 4. Layered microstructures used to derive the Wiener bounds on the effective dielectric permeability. The electric field is denoted by E.

Similarly when the field is oriented perpendicularly to the layers we use the formula for capacitors connected in series and obtain

$$\bar{\varepsilon}_{B2}^{-1} = f_A\, \varepsilon_A^{-1} + f_B\, \varepsilon_B^{-1} \ . \tag{19}$$

Equation (18) corresponds to no screening of the field while Eq. (19) corresponds to maximum screening. The rigorous bounds on $\bar{\varepsilon}$ described by these equations were first obtained by Wiener.[15] If ε_A and ε_B are real Eqs. (18) and (19) describe the bounds on $\bar{\varepsilon}$ when ε_A, ε_B, f_A and f_B are known.

However we are mainly interested in the more general case of complex dielectric permeabilities. Here the bounds can be depicted as lines in the complex $\bar{\varepsilon}$-plane enclosing the region of allowed values of $\bar{\varepsilon}$. Methods for deriving these bounds have been given by Bergman[17,35] and Milton.[18,36,37] The bounds are obtained from the analytical properties of $\bar{\varepsilon}$ as a function of the permeabilities of the components as given by Bergman.[38] Alternatively, variational principles[16,37] can also be used. The Wiener bounds, Eqs. (18) and (19), hold for the case of complex $\bar{\varepsilon}$, irrespective of microstructure. Only ε_A and ε_B need to be known. As f_A and $f_B = 1 - f_A$ are varied Eq. (18) traces out a straight line between ε_A and ε_B, while Eq. (19) becomes a circular arc joining ε_A and ε_B. An example of these bounds is given in Fig. 5.

Now we incorporate gradually more structural information in the bounds on the effective dielectric permeability. First the case when the volume fractions f_A and f_B are known in addition to ε_A and ε_B is considered. It can be shown that the bounds take the form[13,16-18]

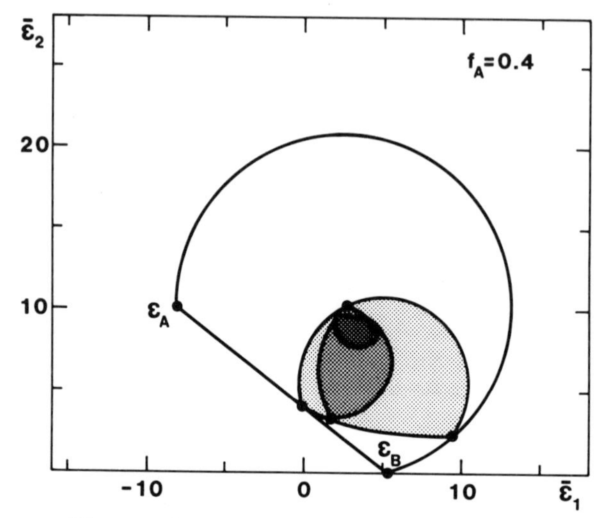

Fig. 5. Rigorous bounds for the effective dielectric permeability of a two-component composite with $\varepsilon_A = -8 + 10i$ and $\varepsilon_B = 5$. These values were chosen to obtain a good display of the features of the bounds. The lines joining the points ε_A and ε_B denote the Wiener bounds $\bar{\varepsilon}_{B1}$ and $\bar{\varepsilon}_{B2}$. The lightly shaded area is enclosed by the bounds for anisotropic composites, $\bar{\varepsilon}_{B3}$ and $\bar{\varepsilon}_{B4}$. Here f_A was set equal to 0.4. The medium shaded area is enclosed by the bounds for isotropic composites, $\bar{\varepsilon}_{B5}$ and $\bar{\varepsilon}_{B6}$. Finally, the heavily shaded area is enclosed by the fourth order bounds, with the structural parameter x equal to 0.1.

$$\bar{\varepsilon}_{B3} = \varepsilon_B + \frac{f_A \varepsilon_B (\varepsilon_A - \varepsilon_B)}{\varepsilon_B + L f_B (\varepsilon_A - \varepsilon_B)} \tag{20}$$

and

$$\bar{\varepsilon}_{B4} = \varepsilon_A + \frac{f_B \varepsilon_A (\varepsilon_B - \varepsilon_A)}{\varepsilon_A + L f_A (\varepsilon_B - \varepsilon_A)}\,, \tag{21}$$

where L is an effective depolarisation factor that can vary between zero and unity. The bounds of Eqs. (20) and (21) become circular arcs in the $\bar{\varepsilon}$ plane joining $\bar{\varepsilon}_{B1}(f_A)$ and $\bar{\varepsilon}_{B2}(f_A)$, as seen in Fig. 5. The extensions of the circular arcs pass through the points ε_A and ε_B, respectively. When ε_A and ε_B are real Eqs. (20) and (21) give bounds for the real $\bar{\varepsilon}$ for the case when f_A, f_B and L are known, as shown first by Hashin and Shtrikman.[16] The area within the bounds corresponds to the effective dielectric permeabilities of anisotropic composites. Such materials have in general three principal $\bar{\varepsilon}$'s related to the three principal directions of the structure.

Even narrower bounds are obtained for the case of isotropic composites. For the case of complex $\bar{\varepsilon}$ they were first derived by Bergman[35] and Milton.[36] Subsequently the bounds were reformulated by Aspnes[13] in a convenient way. In this chapter we employ the latter formulation and write

$$\bar{\varepsilon} = \frac{\varepsilon_A\varepsilon_B + 2\varepsilon_h(f_A\varepsilon_A + f_B\varepsilon_B)}{2\varepsilon_h + f_A\varepsilon_B + f_B\varepsilon_A} \quad . \tag{22}$$

Here ε_h is the dielectric permeability for an imaginary material in which the phases A and B are taken to be embedded. It represents the properties of the average neighbourhood of inclusions of materials A and B. The Bergman-Milton bounds $\bar{\varepsilon}_{B5}$ and $\bar{\varepsilon}_{B6}$ are obtained when

$$\varepsilon_h = x\varepsilon_A + (1\text{-}x)\varepsilon_B \tag{23}$$

and

$$\varepsilon_h^{-1} = x\varepsilon_A^{-1} + (1\text{-}x)\varepsilon_B^{-1} \quad , \tag{24}$$

respectively. Here x is a structural parameter which can take values between zero and unity and is related to the pair and three-point correlation functions of the material. The evaluation of this parameter is discussed in Sec. IV. When x is varied, the bounds $\bar{\varepsilon}_{B5}$ and $\bar{\varepsilon}_{B6}$ form two circular arcs, which join the points representing $\bar{\varepsilon}_{B3}$ and $\bar{\varepsilon}_{B4}$ for L = 1/3, as shown in Fig. 5. These arcs enclose the possible $\bar{\varepsilon}$'s for isotropic composites. The extensions of the arcs pass through $\bar{\varepsilon}_{B1}(f_A)$ and $\bar{\varepsilon}_{B2}(f_A)$, respectively. Bergman[17] has shown that the bound given by Eqs. (22) and (24) is not attainable and can be somewhat improved.

By incorporating more structural information, i.e., higher-order correlation functions, into the theory a whole hierarchy of narrower bounds can be obtained, as shown by Milton[18] and Milton and McPhedran.[37,39] We return to this point in Sec. IV where we discuss bounds for the case when the parameter x is known. Recently the theory of rigorous bounds on the complex $\bar{\varepsilon}$ has also been applied to materials with more than two components.[40-42]

In the rest of this section we will consider various theories for the effective dielectric permeability that hold in special cases. These so called effective medium theories are compared to the bounds outlined above. We now specify some explicit microstructures and carry out calculations of the effective $\bar{\varepsilon}$. Even if strictly valid only for one microstructure these effective medium expressions are often good approximations for many materials encountered in practise. The microstructure is represented with Random Unit Cell (RUC) models which are simple enough to permit a theoretical treatment and yet do not leave out the essential physics. Figure 6a, b shows two cases which are regarded as typical: a separated-grain structure, with particles of "A" embedded in a continuous host of "B", and an aggregate structure in which "A" and "B" enter on an equal footing to form a space-filling random mixture.

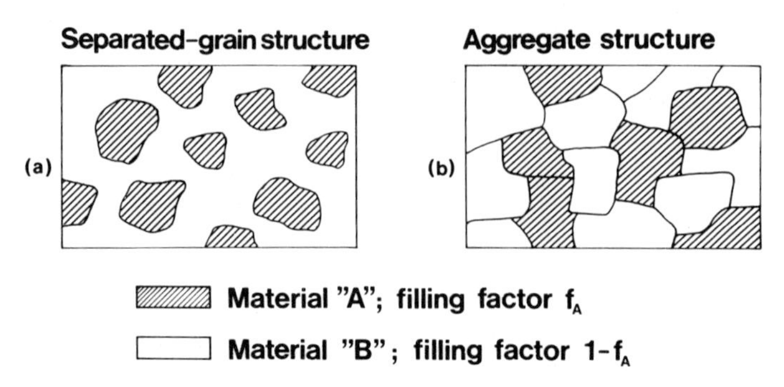

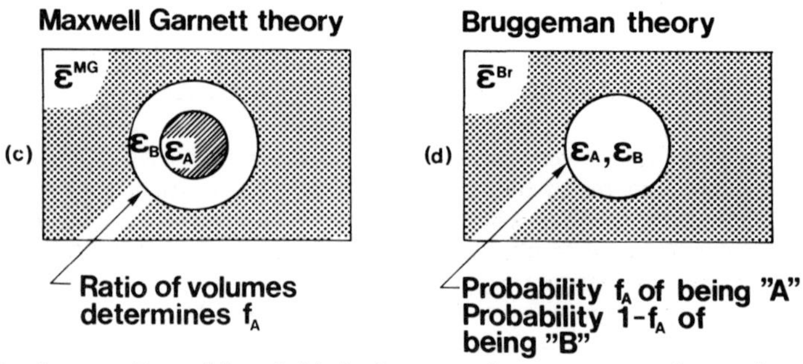

Fig. 6. Parts (a) and (b) depict two microstructures of two-phase composite materials. Parts (c) and (d) show the corresponding RUC's used to derive the Maxwell Garnett and Bruggeman EMT's. The RUC's are embedded in the effective media. (From Ref. 7).

RUC's are now defined[7,43-45] so as to account for the major features of the microstructures. The RUC is taken to be embedded in an effective medium, whose properties are as yet undefined. For the separated grain structure, the RUC is a core of "A" surrounded by a concentric shell of "B", as depicted in Fig. 6c. The ratio of the core volume to the shell volume is equal to f_A. For the aggregate structure, the inherent structural equivalence of the components is ensured by letting the RUC have a probability f_A of being "A" and a probability f_B of being "B", as shown in Fig. 6d. The RUC is spherical or nonspherical depending on the shape of a "typical" particle.

The basic definition of an effective medium is that the RUC, when embedded in the effective medium, should not be detectable in an experiment using electromagnetic radiation confined to a specific wavelength range. In other words, the extinction of the RUC should be the same as if it were replaced with a material characterized by $\bar{\varepsilon}$. This criterion makes it fruitful to use a recently derived[46] "optical theorem" for absorbing media; it relates the extinction of the spherical cell compared to that of the surrounding medium, C_{ext}, with the scattering amplitude in the direction of the impinging beam, S(0), by

$$C_{ext} = 4\,\pi\,\mathrm{Re}\,[\,S(0)/k_e^2]\ , \tag{25}$$

where

$$k_e = 2\,\pi\,\bar{\varepsilon}^{1/2}/\lambda \tag{26}$$

denotes the wavevector amplitude in the effective medium. Equation (25) is seen to be a generalization of the usual[47] optical theorem for nonabsorbing media, but in the present case C_{ext} can be either positive or negative. From the definition of an effective medium it now follows that $C_{ext} = 0$, i.e.,

$$S(0) = 0\ , \tag{27}$$

which expresses the fundamental property of an effective medium. This condition has been proposed earlier[48,49] in somewhat different contexts.

The condition of Eq. (27) is very convenient because S(0) can easily be obtained for spheres and coated spheres by the Lorenz-Mie theory.[47,50] Below we mainly derive effective medium theories where the RUC's have a spherical geometry, but we also comment on some results for ellipsoids. From Lorenz-Mie theory S(0) can be written[47,50] as an expansion in electric and magnetic multipoles, or alternatively as a series expansion in $(k_e b)$ where b is the RUC radius.

The Maxwell Garnett (MG) theory[51] corresponds to the RUC in Fig. 6c. Hence we use the series appropriate to a coated sphere (cs) and obtain[50]

$$S^{cs}(0) = i\,(k_e b)^3\,\frac{(\varepsilon_B - \bar{\varepsilon})\,(\varepsilon_A + 2\varepsilon_B) + f_A\,(2\varepsilon_B + \bar{\varepsilon})\,(\varepsilon_A - \varepsilon_B)}{(\varepsilon_B + 2\bar{\varepsilon})\,(\varepsilon_A + 2\varepsilon_B) + f_A\,(2\varepsilon_B - 2\bar{\varepsilon})\,(\varepsilon_A - \varepsilon_B)} + O\,[(k_e b)^5]. \tag{28}$$

The filling factor is

$$f_A = \frac{a^3}{b^3}\ , \tag{29}$$

where a(b) is the radius of the inner (outer) sphere in Fig. 6c. In the small sphere limit, the effective medium condition can be satisfied by setting the leading term in Eq. (28) equal to zero. This yields (with $\bar{\varepsilon} \equiv \bar{\varepsilon}^{MG}$)

$$\frac{\bar{\varepsilon}^{MG} - \varepsilon_B}{\bar{\varepsilon}^{MG} + 2\varepsilon_B} = f_A\,\frac{\varepsilon_A - \varepsilon_B}{\varepsilon_A + 2\varepsilon_B} \tag{30}$$

or, rewritten,

$$\bar{\varepsilon}^{MG} = \varepsilon_B \frac{\varepsilon_A + 2\varepsilon_B + 2f_A(\varepsilon_A - \varepsilon_B)}{\varepsilon_A + 2\varepsilon_B - f_A(\varepsilon_A - \varepsilon_B)} . \tag{31}$$

Equation (30) or (31) is the constitutive formula for the Maxwell Garnett EMT. By making the replacements $A \rightarrow B$ and $B \rightarrow A$ one obtains analogous relations for the inverted structure. Our derivation does not require that f_A be small. However, it is clear that for a sufficiently large filling factor one reaches a point where the detailed particle-particle interactions must be considered explicitly. Obviously, such structural multipole features can not be encompassed by the MG approach, and supplementary information is required. To shed some light on this issue, we note that the Maxwell Garnett theory is in acceptable agreement[52] with the recently derived[53-55] exact theories for cubic arrangements of identical spheres as long as $f_A < 0.4$. For aperiodic arrangements - such as those normally occurring in experimental samples - the structural multipoles are expected[44] to set in at lower filling factors since close approach amongst the spheres is permitted.

It should be noted that the two MG expressions (Eq. 31 with and without the replacements $A \leftrightarrow B$) are equivalent to the Bergman-Milton bounds, Eqs. (22)-(24), when $x = 0$ and $x = 1$, respectively. The MG expressions are situated on the points in the $\bar{\varepsilon}$-plane where the two bounds for isotropic composites cross (see also Fig. 8).

The Maxwell Garnett theory can also be extended to the case of ellipsoidal RUC's. This is not trivial, as discussed in Refs. 7 and 14, but depends on the shape of the shell in the RUC. When the shell in the RUC has the same depolarisation factor, L, as the inner particle a simple result is obtained, namely the bounds $\bar{\varepsilon}_{B3}$ and $\bar{\varepsilon}_{B4}$ given by Eqs. (20) and (21). Hence these bounds are attained by a geometry consisting of aligned ellipsoids. For randomly oriented ellipsoids one has to average over the three principal axes; the result for $\bar{\varepsilon}$ must then be within the bounds for isotropic composites.

The Bruggeman theory[56] is derived from the RUC in Fig. 6d. We use the series expansion for a sphere (s) and obtain[47,50]

$$S^s(0) = i\,(k_e b)^3 \frac{\varepsilon - \bar{\varepsilon}}{\varepsilon + 2\bar{\varepsilon}} + O\,[\,(k_e b)^5]\ , \tag{32}$$

where b is the radius of the RUC in Fig. 6d and ε denotes ε_A or ε_B. Considering again the small sphere limit, it is found that (with $\bar{\varepsilon} \equiv \bar{\varepsilon}^{Br}$)

$$f_A \frac{\varepsilon_A - \bar{\varepsilon}^{Br}}{\varepsilon_A + 2\bar{\varepsilon}^{Br}} + (1 - f_A) \frac{\varepsilon_B - \bar{\varepsilon}^{Br}}{\varepsilon_B + 2\bar{\varepsilon}^{Br}} = 0, \tag{33}$$

where we have invoked the probability f_A for the RUC of having $\varepsilon = \varepsilon_A$ and the probability f_B of having $\varepsilon = \varepsilon_B$. Equation (33) is the constitutive formula for the Bruggeman theory. For high f_A's, the structural multipole effects are expected to enter as for the Maxwell Garnett theory, but no detailed study has yet appeared. Note that Eq. (33) is symmetrical with respect to exchange of the components of the material.

Two other EMT's can be derived from RUC arguments. These are the theories of Ping Sheng[57] (PS) and Bruggeman-Hanai[56,58] (BH). The former is a compositionally symmetrized form of the Maxwell Garnett theory. The RUC, depicted in Fig. 7a, is a coated sphere whose core and shell can be either "A" or "B". The relative occurrence of the two varieties of RUC is determined[57] by counting the number of equally possible configurations corresponding to different positions of the inner sphere in the RUC. When "A" is the core and "B" is the shell, this number is

$$v_1 = (1 - f_A^{1/3})^3 . \tag{34}$$

For the opposite situation we obtain

$$v_2 = (1 - (1 - f_A)^{1/3})^3 . \tag{35}$$

This argument requires that the inner sphere of the RUC be placed eccentrically. We approximate the ensemble of various eccentric structures by RUC's being concentric coated spheres. Taking the small sphere limit of Eq. (28) for each of the two varieties of RUC and setting $\Sigma S^{cs}(0) = 0$, it is found that (with $\bar{\varepsilon} \equiv \bar{\varepsilon}^{PS}$)

$$v_1 \frac{(\varepsilon_B - \bar{\varepsilon}^{PS})(\varepsilon_A + 2\varepsilon_B) + f_A(2\varepsilon_B + \bar{\varepsilon}^{PS})(\varepsilon_A - \varepsilon_B)}{(\varepsilon_B + 2\bar{\varepsilon}^{PS})(\varepsilon_A + 2\varepsilon_B) + 2f_A(\varepsilon_B - \bar{\varepsilon}^{PS})(\varepsilon_A - \varepsilon_B)} +$$

$$+ v_2 \frac{(\varepsilon_A - \bar{\varepsilon}^{PS})(\varepsilon_B + 2\varepsilon_A) + (1 - f_A)(2\varepsilon_A + \bar{\varepsilon}^{PS})(\varepsilon_B - \varepsilon_A)}{(\varepsilon_A + 2\bar{\varepsilon}^{PS})(\varepsilon_B + 2\varepsilon_A) + 2(1 - f_A)(\varepsilon_A - \bar{\varepsilon}^{PS})(\varepsilon_B - \varepsilon_A)} = 0. \tag{36}$$

Equation (36) approaches the Maxwell Garnett results when f_A is close to zero or unity. It should be remarked that Ping Sheng made his derivations[57] for the more general case of spheroidal particles; we have chosen to give here the more tractable formula appropriate to spherical particles.

RANDOM UNIT CELLS

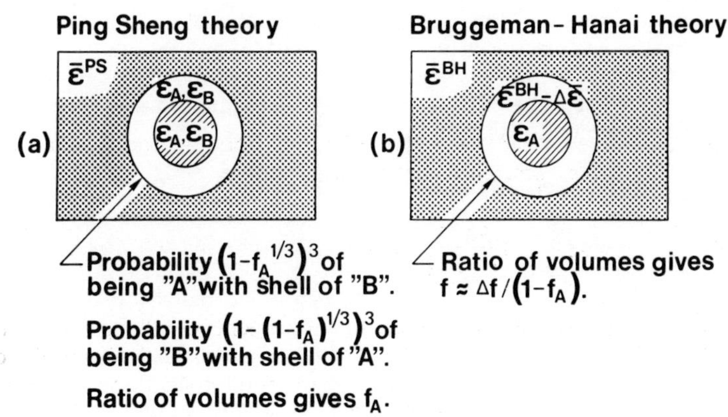

Fig. 7. Parts (a) and (b) show the RUC's used to derive the Ping Sheng and Bruggeman-Hanai EMT's. (From Ref. 7).

The Bruggeman-Hanai theory can be obtained from the RUC in Fig. 7b. It consists of a spherical core of "A" surrounded by a shell whose dielectric permeability is $\bar{\varepsilon}^{BH}$ minus the contribution $\Delta\bar{\varepsilon}$ from the core itself. This is an intermediate situation between those of the Maxwell Garnett and Bruggeman theories. Rewriting Eq. (28) with the notation of Fig. 7b and taking the small sphere limit, it is found that (with $\bar{\varepsilon} \equiv \bar{\varepsilon}^{BH}$)

$$- \Delta\bar{\varepsilon} (\varepsilon_A + 2\bar{\varepsilon}^{BH} - 2\Delta\bar{\varepsilon}) + f' (3\bar{\varepsilon}^{BH} - 2\Delta\bar{\varepsilon}) (\varepsilon_A - \bar{\varepsilon}^{BH} + \Delta\bar{\varepsilon}) = 0 \ , \tag{37}$$

where f' is the ratio of the volume of the inner sphere to that of the outer sphere. The volume fraction of "A" in the RUC is f_A, and the volume fraction in the shell is $f_A - \Delta f$. This yields

$$f' \approx \Delta f/(1 - f_A) \ . \tag{38}$$

It is now straightforward[14] to prove that

$$\frac{\varepsilon_A - \bar{\varepsilon}^{BH}}{\varepsilon_A - \varepsilon_B} = (1 - f_A) \left(\frac{\bar{\varepsilon}^{BH}}{\varepsilon_B}\right)^{1/3} \ . \tag{39}$$

A formula for the inverted structure is obtained by the replacements $A \leftrightarrow B$. Equation (39) was first derived by Bruggeman[56] and has since been studied by Hanai[58] and others.[59,60] The theory is sometimes called the unsymmetrical Bruggeman theory.

Theories which can be derived from spherical RUC's correspond to isotropic structures and should thus fall within the Bergman-Milton bounds $\bar{\varepsilon}_{B5}$ and $\bar{\varepsilon}_{B6}$. In Fig. 8 we show these bounds in a special case together with results from the effective medium theories described above. It is seen that all the theoretical results fall inside the bounds, with the exception of the MG results which fall on the bounds as noted above.

Effective medium theories which are derived from ellipsoidal unit cells correspond in general to anisotropic structures and should fall within the wider bounds $\bar{\varepsilon}_{B3}$ and $\bar{\varepsilon}_{B4}$. Only if a completely random orientation of the ellipsoids in the material is assumed will the effective dielectric permeability fall within the Bergman-Milton bounds for isotropic structures.

The effective medium theories described in this section are often in fairly good agreement with measurements on metal-insulator composite coatings. However, the experimental data show different behaviour depending on the microstructure of the coatings. One type of composite consists of metal particles in an amorphous insulator matrix. Here the Maxwell Garnett theory is usually in good agreement with the experimental $\bar{\varepsilon}$ for low filling factors. This is expected since the composite has a separated-grain structure. However, for $f > 0.2$ to 0.3 discrepancies start to appear. As noted above, at such high filling factors the MG theory is probably inadequate for describing the multipolar interactions between the particles. This type of composite (the "MG type") includes Co-Al_2O_3 (Ref. 7), Au-Al_2O_3 (Refs. 61, 62), Ni-Al_2O_3 (Ref. 6), Ni-SiO_2 and Ni-MgO (Ref. 63) as well as metal-polymer composites (Refs. 64, 65), and probably Ag-SiO_2 and Au-SiO_2 (Ref. 66). The description of the optical properties of "MG type" composites at high filling factors is a complicated problem, although in some cases the Ping Sheng theory works remarkably well.[57,62,64,65] We take a different approach to this problem in Sec. IV below.

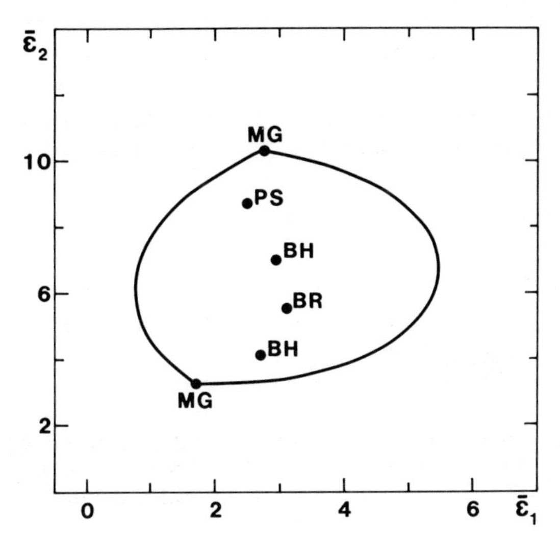

Fig. 8. Bergman-Milton bounds for the effective dielectric permeability of an isotropic two-component composite with $f_A = 0.4$, $\varepsilon_A = -8 + 10i$ and $\varepsilon_B = 5$. These values were also used in Fig. 5. The results of the Maxwell Garnett (MG), Bruggeman (Br), Bruggeman-Hanai (BH) and Ping Sheng (PS) theories are indicated.

Another type of composite consists of a mixture of metal and crystalline insulator grains. This structure is similar to the aggregate model in Fig. 6b, so we call it the "Br type" composite structure. Indeed, the optical properties can be well described by the Bruggeman theory up to quite high ($f \sim 0.3$ to 0.4) filling factors. Examples of "Br type" composites are Au-MgO (Refs. 6, 61, 67-69) and Ag-MgO. The Bruggeman theory predicts a metal-insulator transition when $f_{metal} = 1/3$, but in experiments the transition is often observed at higher filling factors. A phenomenological extension of the Bruggeman theory for the description of the optical properties near the metal-insulator transition has been worked out by Berthier et al.[70] This theory, as well as a renormalization group calculation,[71] show the qualitative features seen in experiments.

To sum up, in this section we have gradually incorporated more and more structural information into the derivation of rigorous bounds on $\bar{\varepsilon}$ and effective medium theories. The Wiener bounds, $\bar{\varepsilon}_{B1}$ and $\bar{\varepsilon}_{B2}$, which are valid when only the dielectric permeability of the constituents are known, are attained for layer structures. These expressions have recently been found very useful for describing the optical properties of layered superlattices.[72] When the filling factors of the two components are also known we obtain the wider bounds $\bar{\varepsilon}_{B3}$ and $\bar{\varepsilon}_{B4}$ which hold for anisotropic composites. They are attained by the Maxwell Garnett theory for oriented ellipsoids. For isotropic composites the Bergman-Milton bounds $\bar{\varepsilon}_{B5}$ and $\bar{\varepsilon}_{B6}$ are found. The two points where the boundary circles intersect correspond to the Maxwell Garnett theory for spheres. The bound $\bar{\varepsilon}_{B5}$ is attained by composites where both phases percolate[17] as seen from Eq. (23), while $\bar{\varepsilon}_{B6}$ cannot be attained by real composites.[17]

In the next section we will incorporate additional structural information into the derivation of rigorous bounds and effective medium theories by use of a structural parameter related to the pair and three-point correlation functions of the material.

IV. EFFECTS OF PAIR AND THREE-POINT CORRELATION FUNCTIONS

The structure of a composite material can be completely specified by various n-point correlation functions. Recently a very general approach to this problem has been published by Torquato[73] and Torquato and Stell.[74] They define general correlation functions and show how different descriptions are related to one another. In this section we will only consider the correlation functions of the lowest orders, namely the pair and three-point correlation functions of particles dispersed in a matrix. This geometry is similar to the separated-grain structure in Fig. 6a. We also briefly review correlation effects in the class of cell materials, which corresponds to the aggregate structure in Fig. 6b. From the correlation functions more narrow bounds on $\bar{\varepsilon}$ than those considered in Sec. III can be obtained. Some novel effective medium theories which take into account information on the pair and three-point correlation functions are also described.

We first consider a structure consisting of spheres dispersed in a continuous matrix. In order to obtain more restrictive bounds than the Bergman-Milton bounds for isotropic composites, we must evaluate the parameter x which appears in Eqs. (23) and (24). The resulting bounds we denote the fourth-order bounds. An example of these bounds for $x = 0.1$ is given by the darkest area in Fig. 5. The fourth-order bounds were first described by Beran.[75]

The structural parameter $x = x_2 + x_3$ is related to the pair correlation function, $g_2(r)$, and the three-point correlation function, $g_3(r_{12}, r_{13}, r_{23})$, by the expressions[24,76]

$$x_2 = \frac{3f_A}{2(1-f_A)} \sum_{l=1}^{\infty} l(l+1) \int_{2a}^{\infty} g_2(r) \frac{a^{2l+1}}{r^{2l+2}} \, dr, \tag{40}$$

where a is the particle radius. Furthermore[24, 76]

$$x_3 = \frac{9f_A{}^2}{32\pi^2 (1-f_A)} \sum_{l=1}^{\infty} l(l+1) \int\int d\bar{r}_{12}\, d\bar{r}_{23}\, [g_3 (r_{12}, r_{13}, r_{23}) -$$

$$- g_2 (r_{12})\, g_2 (r_{23})]\, P_{l+1} (\hat{r}_{21} \cdot \hat{r}_{23}) \frac{a^{2l+2}}{r_{12}{}^{l+2}\, r_{23}{}^{l+2}}, \qquad (41)$$

where $\bar{r}_{mn}$ and $\hat{r}_{mn}$ are vector distances between particles m and n and unit vectors in direction m to n, respectively. The three-point correlation function is denoted by g_3, P_{l+1} are Legendre polynomials, and r_{mn} are the absolute values of the particle-particle distances. Note that Refs. 24 and 76 use a parameter K which is related to x by $K = 2f_A (1-f_A) x/9$.

The parameter x has been evaluated for some simple model structures. A particularly simple case is a fractal structure[77] where the pair correlation function is given by[77,78]

$$g_2 (r) = \begin{cases} (\xi/r)^{3-D} & 2a<r<\xi \\ 1 & r>\xi. \end{cases} \qquad (42)$$

Here D is the fractal dimension and ξ is the correlation length of the structure. The correlation length signifies the upper cutoff and $2a$ the lower cutoff to the range over which the structure is fractal. The filling factor of the structure is related to the upper and lower cutoffs by

$$f = L^* \left(\frac{a}{\xi}\right)^{3-D}, \qquad (43)$$

where L^* is a measure of the "holes" in the structure and related to the lacunarity.[77] It can take values between zero and unity. A similar relation has been used to determine the porosity of sandstones.[79] When the particles are touching, the parameter L^* is equal to unity. It will be seen that this case gives the highest value of x.

We first consider the integral over the pair correlation function, x_2. Inserting Eq. (42) in Eq. (40), one can immediately evaluate the integral to obtain

$$x_2 = \frac{3f}{2(1-f)} \sum_{l=1}^{\infty} \frac{l(l+1)}{(2l-D+4)\, 2^{2l+1}} \left(\frac{\xi}{2a}\right)^{3-D} =$$

$$= \frac{3L^*}{2(1-f)} \sum_{l=1}^{\infty} \frac{l(l+1)}{(2l-D+4)\, 2^{2l-D+4}}, \qquad (44)$$

where we have used Eq. (43). The value of x_2 is only weakly dependent on f for $f < 0.1$.

The parameter x_3 has to be evaluated numerically. We evaluated Eq. (41) using the method of Felderhof.[76] The calculation was performed for a model three-point correlation function valid for at least some fractal structures, namely[23]

$$g_3(r_{12}, r_{13}, r_{23}) - g_2(r_{12})\, g_2(r_{23}) =$$

$$= 2.5 + \xi^{6-2D}\, [(r_{23}\, r_{13})^{D-3} + (r_{13}\, r_{12})^{D-3} - (r_{12}\, r_{23})^{D-3}]\, /2. \qquad (45)$$

In Fig. 9, x_2 and x_3 are depicted as a function of fractal dimension when $L^* = 1$ and the filling factor goes to zero. It is seen that the effect of three-point correlations is small for $D < 2$, but it becomes more important for higher fractal dimensions. Physically it is expected that x should increase towards unity as D comes close to three. It should be noted that quite different mean-field approaches to the optical properties of fractal clusters have also been proposed.[80]

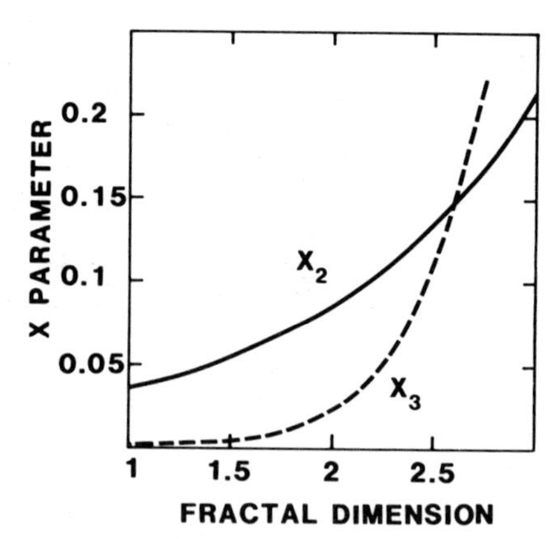

Fig. 9. Structural parameters x_2 and x_3 as a function of fractal dimension in the limit of low filling factor. The calculations were performed for fractal structures with $L^* = 1$.

Recently the information available on x has increased substantially and accurate computations have been performed for random distributions of spheres dispersed in a matrix. The case of impenetrable or hard spheres was treated by Felderhof[26] and by Torquato and Lado.[24] Improved values of x as a function of filling factor were later computed by Beasley and Torquato.[25] Another simple model considers random distributions of spheres that are allowed to penetrate one another. In this case values of x were computed by Torquato and Stell.[27,81]

The case of spheres with an arbitrary degree of penetrability has also been thoroughly studied.[27,82] For example one may consider spheres with a hard core and a penetrable shell. The spheres are taken to be randomly distributed but subject to the condition that they may not penetrate each others' cores. The penetrable sphere model (PSM) and impenetrable sphere model (ISM) are obtained in the limits of no core and no shell, respectively. The ISM can be considered as a generalization of the Maxwell Garnett theory in order to include pair and three-point correlation effects. The PSM, on the other hand, corresponds to a quite different geometry in that a higher connectivity between the particles is obtained. It may be a viable model for sintered or partially sintered composites.

The computation of structural parameters of random distributions of spheres is rather involved and a detailed exposition falls outside the scope of this chapter. In Fig. 10 literature data[25,27,81] on the structural parameter x is shown as a function of filling factor for the ISM and PSM models. As expected the PSM shows a greater departure from the Maxwell Garnett theory (x=0) than the ISM.

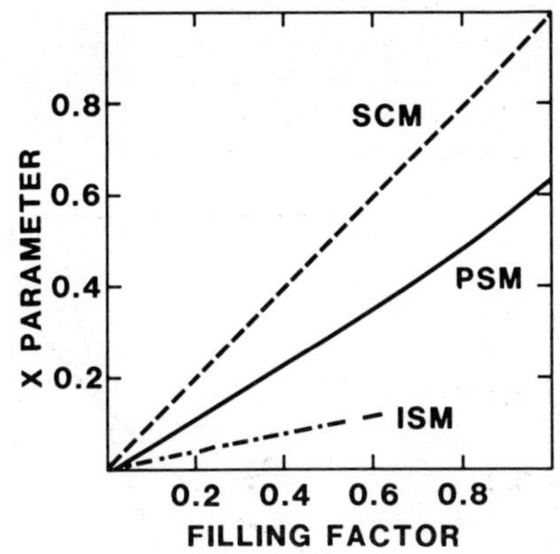

Fig. 10. Structural parameter x as a function of filling factor for spherical cell materials (SCM) as well as for dispersions of penetrable spheres (PSM) (Refs. 27, 81) and impenetrable spheres (ISM) (Ref. 25).

So far we have only treated generalizations of the separated-grain structure depicted in Fig. 6a. We also make some comments on the structural parameter x for composites having an aggregate structure as shown in Fig. 6b. Such composites are also called cell materials, a concept introduced by Miller.[83] A cell material is defined by the following requirements:[83] (1) space is completely filled with cells; (2) the cells are distributed so that the material is statistically homogeneous and isotropic; and (3) the material or property of a cell is statistically independent of the other cells. The fourth-order bounds for cell materials were derived by Miller[83] and further analyzed by McPhedran and Milton.[39] For cell materials composed of randomly oriented spheroidal cells with depolarization factors L_1, L_2 and L_3 they obtained the following expression[39] for the structural parameter x:

$$x = (1 - f_A)\, \rho + f_A\, (1 - \rho) \ , \tag{46}$$

where

$$\rho = [3\,(L_1{}^2 + L_2{}^2 + L_3{}^2) - 1]\ /2. \tag{47}$$

For the case of spherical cells Eq. (47) reduces to $\rho = 0$, which yields $x = f_A$. Obviously, composites described by the Bruggeman theory have the same value of x as a spherical cell material. From Fig. 10 we see that dispersions of impenetrable spheres have properties quite close to the Maxwell Garnett theory ($x=0$), while dispersions of penetrable spheres should have properties intermediate between those of the Maxwell Garnett and Bruggeman theories.

Recently there has been an increasing interest in the formulation of effective medium theories taking into account correlation effects. Various schemes have been devised for improving the Maxwell Garnett theory by including dipole-dipole and higher multipole interactions between the particles. Random unit cells containing many particles[44,84] and the effect of topological disorder in the lattice gas model[85] have been studied. Effects of the pair correlation function in effective medium expressions is another subject of current interest.[86-88] A thorough study of dipole and multipole interactions in dilute dispersions of spheres was carried out by Felderhof and Jones.[89] However, because the pair correlation and three-point correlation functions both enter into the calculation of the fundamental structural parameter x, we believe that any effective medium theory which is to be valid over a range of filling factors must include both pair and three-point contributions. Theories including just the pair correlation function should be good only for low filling factors. Effective medium theories which depend explicitly on the value of x have only recently begun to appear.[23,90,91] We conclude this section with two examples of such theories and apply them to relevant experimental situations.

First we consider gas evaporated metal particle coatings. They are produced by evaporation of metal in a few Torr of an inert gas, sometimes with a small addition of oxygen.[92] This process leads to large metal particle aggregates with a fractal structure.[23,93] When collected on a substrate, the aggregates are connected over macroscopic distances. The deposits are sooty, loosely packed powders, where the volume fraction of particles is typically less than a few percent. The coatings have a structure where both phases percolate, i.e., the particle phase and the surrounding air. This means that the effective medium theory given by Eqs. (22) and (23) should be a good approximation for the optical properties of the coatings.[23] The fractal dimensions of nonmagnetic gas evaporated aggregates[94] are typically in the range 1.9 to 2. Thus a value of x of about 0.09 to 0.10 should be used in the theoretical calculation.

We now apply this theory to the case where each metal particle is coated by an oxide shell. This is introduced by performing the evaporation with oxygen mixed in the gas present during the process. For the dielectric permeability of the particle phase we use the formula for a coated sphere which is[47]

$$\varepsilon_A = \varepsilon_c \frac{\varepsilon_M + 2\varepsilon_c + 2\Omega\,(\varepsilon_M - \varepsilon_c)}{\varepsilon_M + 2\varepsilon_c - \Omega\,(\varepsilon_M - \varepsilon_c)}\,, \tag{48}$$

where ε_M is the dielectric permeability of the metal core, ε_c is that of the oxide coating and Ω is the volume fraction of metal in the particle. It should be kept in mind that Eq. (48) takes into account only the dipole terms in the interaction between core and coating. When the sizes of the metal particles are less than the mean free path of the conduction electrons in the pertinent metal, ε_M must be corrected for boundary scattering of conduction electrons in the particles.[95,96]

In Fig. 11 experimental and theoretical transmittance for a layer of gas evaporated oxide coated aluminium particles is shown. We have used the theory presented above to evaluate the effective dielectric permeability of the layers, and obtained the transmittance of the layers (divided by the transmittance of the substrates) from the standard Fresnel relations in Sec. II. As input data we used the experimentally determined dielectric permeability of aluminium[97,98] and evaporated Al_2O_3.[98-100] The experimentally determined value for the diameter of the aluminium core (~ 2 nm), the filling factor f ~ 0.01 and a value for the weight per unit area 0.15 g/m^2, which is within the error bars of the experimental value, were also used.

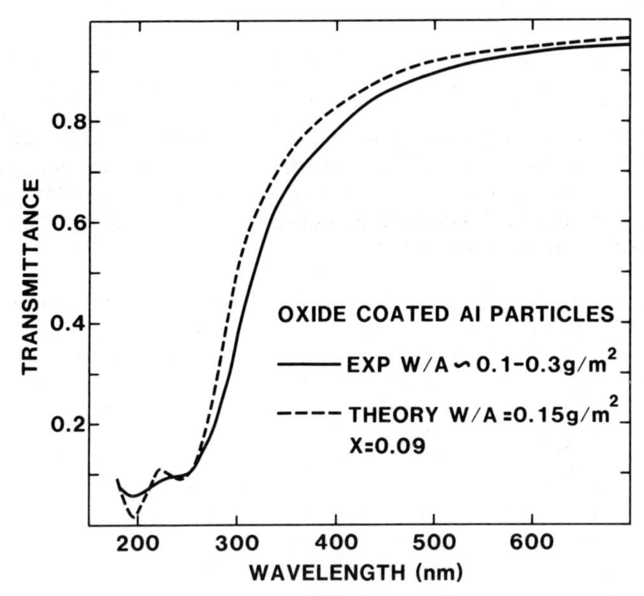

Fig. 11. Transmittance as a function of wavelength for oxide-coated aluminium particles produced by gas evaporation. The full curve denotes experimental results, and the dashed curve denotes calculated values. The filling factor for the particles was 0.01, the diameter of the aluminium core was 2 nm and the volume fraction, Ω, of aluminium in the particles was 0.25. The weight per unit area in the calculations was 0.15 g/m^2, which is within the error bars of the experimental value as shown in the inset. Calculations were carried out using the EMT for fractal structures with structural parameter x = 0.09. (From Ref. 23).

The agreement between theory and experiment is satisfactory. The experimental curve shows a prominent absorption peak at $\lambda \sim 190$ nm with a shoulder on the long wavelength side. These features are present in all gas evaporated oxide coated aluminium samples that we have studied. The theoretical curve shows a sharper structure with peaks at 190 nm and 240 nm. The absorption peak at $\lambda \sim 190$ nm is due to the localized plasma resonance of the conduction electrons in independent aluminium particles, while the peak at $\lambda \sim 240$ nm is a consequence of the percolation structure.

A long standing problem in the field of optical properties of inhomogeneous materials is the applicability of effective medium descriptions in the infrared and far infrared wavelength ranges. In particular an enhanced absorption of small metal particles has been found in the far infrared,[101] which has thrown some doubt on the applicability of current theories. The good agreement between theory and experiment in Fig. 11 prompted us to study gas evaporated oxide coated aluminium particles also in the far infrared.[102] Gas evaporated metal particles display a strikingly high absorption in this range[102, 103] and are convenient for such studies. The samples used for far infrared measurements[101,102] had particles with a median diameter of ~ 8.4 nm, the filling factor was ~ 0.015, and the fraction of unoxidized aluminium in the particles was estimated[92] to be $\Omega \sim 0.35$. In our computations we used the dielectric permeabilities of aluminium[98] and Al_2O_3.[99] The ε's for aluminium were corrected for boundary scattering of conduction electrons in the metal cores. In Fig. 12 we show the Bergman-Milton bounds for a wavelength of 200 μm, the experimental value[101,102] of $\bar{\varepsilon}$ with error bars and the theoretical value for a fractal structure with both phases percolating and x=0.1. It is seen that there is about a factor of 2 difference between theory and experiment. Thus most of the previously reported discrepancies in the far infrared absorption of metal particles can probably be ascribed to structural effects. This conclusion has also been reached by others.[104] An assessment of the remaining discrepancy between theory and experiment requires more detailed experiments with accurate structural characterisation.

Our second example concerns $Co\text{-}Al_2O_3$ coatings[7,91] which are efficient selective absorbers of solar energy. These coatings were produced by electron beam coevaporation from two sources, and consist of 10-30 Å cobalt particles dispersed in an amorphous matrix of mixed cobalt-aluminium-oxide.[7] The particles are to a good approximation dispersed randomly in the oxide and are preferentially surrounded by an oxide shell[91] for filling factors up to 0.6-0.7. We model this material as a system of randomly dispersed impenetrable oxide-coated cobalt spheres, dispersed in a matrix of the same oxide. The dielectric permeability of the particles is given by Eq. (48), where Ω was estimated[91] to be ~ 0.75 by equating the observed[91] conductivity threshold at $f \sim 0.15$ to 0.20 with the percolation threshold for oxide-coated spheres. For filling factors above the conductivity threshold, both phases are probably percolating and Eqs. (22) and (23) can be used with x given by the values for impenetrable spheres[25] in Fig. 10. We call this scheme the Random Hard Coated Spheres (RHCS) theory.

When the filling factor is less than 0.15, the Maxwell Garnett theory is able to produce good agreement with experiments. In fact, for such low filling factors the MG and RHCS theories are not much different. Fig. 13 shows the imaginary part of the dielectric permeability for Co-Al_2O_3 coatings with different filling factors. It is seen that for f = 0.17 and 0.23 the RHCS theory gives a significantly better fit to experimental values than the MG theory. For f = 0.29 discrepancies also start to appear between the RHCS theory and the measurements. In summary, our knowledge of the parameter x has allowed us to derive a new effective medium theory for Co-Al_2O_3 composites. This theory is accurate over a larger range of filling factors than the previously used MG theory.

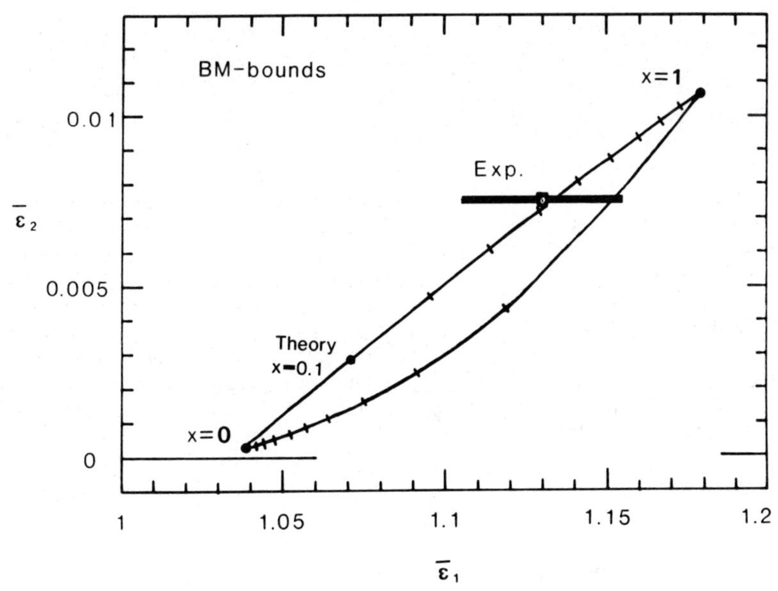

Fig. 12. Bergman-Milton bounds for an isotropic dispersion of Al_2O_3-coated aluminium spheres in air. The computation refers to λ = 200 μm, f_A = 0.015 and Ω = 0.35. The marks on the bounds indicate values pertinent to the structural parameter x going from zero to unity by steps of 0.1. The experimental value with error bars for ε_1 and the theoretical value calculated from the effective medium theory for fractal structures are also shown. (From Ref. 102).

V. MATERIALS WITH LARGE INHOMOGENEITIES

So far we have only considered two-component materials, where the size of the inclusions of the components are much smaller than the wavelength of light. This allowed us to treat the optical properties of the composites in the quasistatic approximation. In this section we go beyond this approximation and treat absorption and scattering by composite media where the inhomogeneities are comparable to or larger than the wavelength.

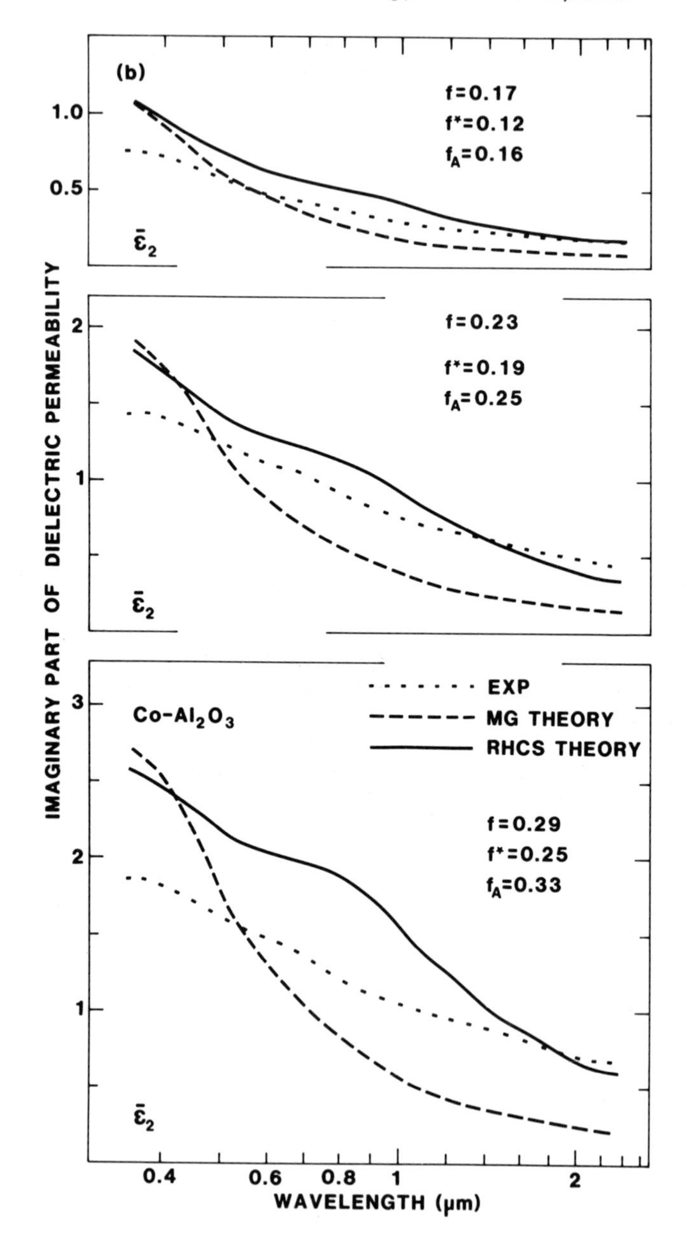
(b)
f=0.17
f*=0.12
f_A=0.16
1.0
0.5
$\bar{\varepsilon}_2$
f=0.23
f*=0.19
f_A=0.25
2
1
$\bar{\varepsilon}_2$
IMAGINARY PART OF DIELECTRIC PERMEABILITY
Co-Al_2O_3
EXP
MG THEORY
RHCS THEORY
f=0.29
f*=0.25
f_A=0.33
3
2
1
$\bar{\varepsilon}_2$
0.4
0.6
0.8
1
2
WAVELENGTH (μm)

Fig. 13. Imaginary part of the effective dielectric permeability as a function of wavelength for Co-Al_2O_3 coatings. Dotted curves denote experimental results, dashed curves denote computations by the Maxwell Garnett theory and full lines denote computations by the Random Hard Coated Spheres theory. The filling factors of cobalt, f, of cobalt particles, f*, and of oxide coated cobalt particles, f_A, are given in the inset. The volume fraction of metal in the oxide coated particles was put equal to 0.75. (From Ref. 91).

Our definition of an effective medium required that the Random Unit Cell, when embedded in the effective medium, should not be detectable by electromagnetic radiation. This led to the criterion that the scattering amplitude S(0) should be zero. However, one could just as well demand that the backscattering amplitude S(π) should be zero. Thus a more rigorous definition of an effective medium is

$$S(0) = S(\pi) = 0, \tag{49}$$

where the scattering is from a RUC embedded in the effective medium. Equation (49) is satisfied as long as only the electric dipole term in the Lorenz-Mie expansion is taken into account, or, equivalently, as long as the quasistatic approximation holds.

The limits of validity of the simple effective medium theories treated in Sec. III can be assessed by computing the contribution of the higher order terms of the Lorenz-Mie expansion to the effective dielectric permeability.[7,14] These large-size limits depend on the filling factor, the wavelength and on the dielectric permeabilities of the constituent materials, and no general rules can be given. However, for the case of metal-insulator composites it appears that the diameter of the metal particles should be less than ~ λ/20 in order that the effective medium theories remain accurate.[7,14]

When the particle size is so large that Eq. (49) is no longer valid one has to resort to extended effective medium theories, as shown by Bohren.[19] These theories employ two effective quantities, namely the effective dielectric permeability discussed in previous sections and an effective magnetic permeability, $\bar{\mu}$. Rigorous results for $\bar{\varepsilon}$ and $\bar{\mu}$ are known only in the limit of small filling factors where we obtain[19]

$$(\bar{\varepsilon}\bar{\mu})^{1/2} = 1 - i\frac{3f_A}{2\,(k_e a)^3}\,S(0) \tag{50}$$

and

$$\bar{\varepsilon}^{1/2} - \bar{\mu}^{1/2} = -i\frac{3f_A}{2\,(k_e a)^3}\,S(\pi)\ . \tag{51}$$

Here we have assumed spherical particles of radius a. From these equations expressions for $\bar{\varepsilon}$ and $\bar{\mu}$ may be obtained. The extension of the theory to higher filling factors remains an object for future research. Also Eqs. (50) and (51) were derived by assuming normally incident light onto a slab of the composite, and their applicability at oblique incidence is still in question.[19]

The effective dielectric and magnetic permeabilities obtained from extended effective medium theories can now be inserted into the expressions for thin film optics in Sec. II. In this way the transmittance and reflectance can be calculated, at least for normally incident light. The extended effective medium theories have so far not been used very much. A detailed treatment of the long-wavelength limit was given in Ref. 105. The main area of application concerns the far infrared absorption of metal particles, where in many cases the magnetic dipole terms are very important.[49,103,106-107] However, the fact that $\bar{\varepsilon}$ and $\bar{\mu}$ are independent quantities has seldom been appreciated; in many cases only the product $\bar{\varepsilon} \cdot \bar{\mu}$ has been studied.

The problem of obtaining rigorous bounds to the effective physical properties of a composite beyond the quasistatic limit has been little explored. The only work has been a generalization of the Wiener bounds to materials with inhomogeneities comparable to the wavelength.[108] However, no bounds on the effective magnetic permeability were given. The treatment of Aspnes[108] has been applied to experiments on pressed powders.[108,109]

A major shortcoming of extended effective medium theories is that they cannot distinguish between scattering and absorption.[19,105] In reality, for materials where the component particles are sufficiently large, diffuse scattering of light plays a major role. In addition to the directly transmitted and specularly reflected light from a slab of the composite, significant amounts of diffusely reflected and transmitted light can be present also. This situation demands a thoroughly different theoretical approach and even the formulas of thin film optics in Sec. II can no longer be used. In order to treat scattering media or materials where both scattering and absorption are important we have to consider multiple scattering or radiative transfer theory.[20-22] In the general case this theory is very complex and results have to be obtained by numerical calculations. A very useful tabulation of results from multiple scattering theory has been published by van de Hulst.[20] If information on the angular distribution of the scattered light is not necessary, important simplifications can be made. This situation is present when measurements of the optical properties of composite materials are performed by an integrating sphere. In this case the specular and diffuse components of the reflectance, and the direct and diffuse transmittance are easily measured. These four quantities can be calculated by a simplified multiple scattering theory, the so called four-flux theory.[21,110-112] An analytical solution of this theory has recently been given by Maheu et al.[112]

Consider now a slab of thickness d which contains a dilute dispersion of particles. In the four-flux theory the radiation field is modeled as consisting of four parts: a collimated beam of intensity $I_c(z)$ and a diffuse beam of intensity $I_d(z)$ propagating to negative z, and a collimated beam $J_c(z)$ and a diffuse beam $J_d(z)$ propagating to positive z. The geometry is shown in Fig. 14. The differential equations for the intensities are[112]

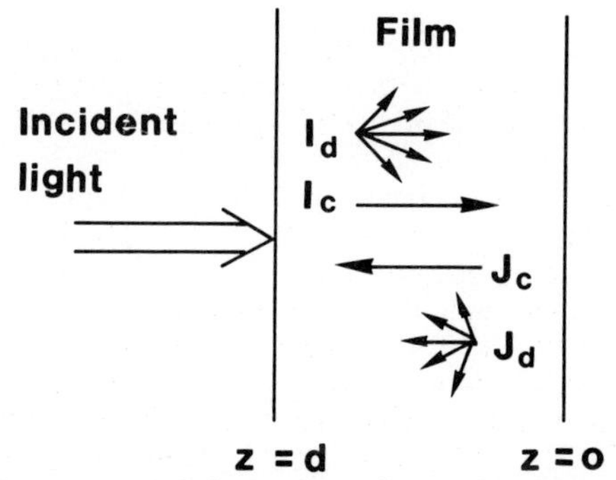

Fig. 14. Geometry and the various fluxes that appear in the four-flux theory. (From Ref. 114).

$$dI_c/dz = (k+s)I_c, \tag{52}$$

$$dJ_c/dz = -(k+s)J_c, \tag{53}$$

$$dI_d/dz = \eta k I_d + \eta(1-\rho_d)sI_d - \eta(1-\rho_d)sJ_d - \rho_c sI_c - (1-\rho_c)sJ_c, \tag{54}$$

$$dJ_d/dz = -\eta k J_d - \eta(1-\rho_d)sJ_d + \eta(1-\rho_d)sI_d + (1-\rho_c)sI_c + \rho_c sJ_c. \tag{55}$$

Here k is the absorption coefficient, s is the scattering coefficient, ρ_c and ρ_d are the ratios of forward to total scattering for the collimated and diffuse beams, respectively, and η is a parameter that determines the average path length travelled by the diffuse beams as compared to the collimated ones. As discussed by Maheu et al.[112] the coefficients k, s and ρ_c can be obtained from single scattering calculations by Lorenz-Mie theory for the case of spherical particles. For the other parameters, rigorous results are known only for extreme cases. The parameter η is unity for a collimated beam and two for a semi-isotropic diffuse beam.[112] The forward scattering ratio ρ_d has been calculated by Ishimaru[21] for the case of isotropic incident light. The other extreme case, i.e., very anisotropic incident light, would lead to $\rho_c = \rho_d$, as assumed by Maheu et al.[112]

The set of equations (52)-(55) can be solved in order to obtain the transmittance and reflectance for a slab of the composite material. The calculations are elementary, although rather involved, so that only the results of the calculations are presented below.[112] The total transmittance is the sum of the direct component T_c and the diffuse component T_d, i.e.

$$T = T_c + T_d, \tag{56}$$

where[112]

$$T_c = \frac{(1-r_c)^2 \exp(-(k+s)\,d)}{1-r_c^2 \exp(-2(k+s)\,d)} \tag{57}$$

and

$$T_d = \frac{(1-r_c)\,(1-r_d)\,\exp\,(-(k+s)\,d)}{(A_1 - (k+s)^2)\,(1-r_c{}^2\,\exp\,(-2(k+s)\,d))}\,\frac{N}{D}\,. \tag{58}$$

Here r_c and r_d are the reflection coefficients at the slab boundaries for collimated and internal diffuse incident radiation, respectively. The parameters in Eq. (58) are given by[112]

$$N = A_1\,[r_dA_3 - A_2 + r_c\,(r_dA_2\text{-}A_3)]\,\cosh\,(A_1d) +$$

$$+\ [(A_5\text{-}r_dA_4)\,(A_3\text{+}A_2r_c) - (A_4\text{-}r_dA_5)\,(A_2\text{+}A_3r_c)]$$

$$\cdot\,\sinh\,(A_1d) + A_1\,[(A_2\text{-}r_dA_3)\,\exp\,[(k+s)d] +$$

$$+\ r_c\,(A_3\text{-}r_dA_2)\,\exp\,(-(k+s)d)], \tag{59}$$

$$D = A_1\,(r_d{}^2\text{-}1)\,\cosh\,(A_1d) + [r_d\,(A_5\text{-}r_dA_4) + r_dA_5 - A_4\,]\,\sinh\,(A_1d), \tag{60}$$

$$A_1 = \eta\,[k^2 + 2(1\text{-}\rho_d)\,ks]^{1/2}, \tag{61}$$

$$A_2 = s\,[\eta k\rho_c + \eta s(1\text{-}\rho_d) + \rho_c(k+s)], \tag{62}$$

$$A_3 = s\,(1\text{-}\rho_c)\,(k+s)\,(\eta\text{-}1), \tag{63}$$

$$A_4 = \eta\,(k + (1\text{-}\rho_d)\,s), \tag{64}$$

$$A_5 = \eta\,(1\text{-}\rho_d)\,s. \tag{65}$$

Similarly the total reflectance is the sum of the specular component R_c and the diffuse component R_d, i.e.

$$R = R_c + R_d. \tag{66}$$

The components are given by

$$R_c = r_c + \frac{r_c\,(1\text{-}r_c)^2\,\exp\,(-2(k+s)\,d)}{1\text{-}r_c{}^2\,\exp\,(-2(k+s)\,d)} \tag{67}$$

and

$$R_d = \frac{(1-r_d)(1-r_c)\exp(-(k+s)d)}{[A_1 - (k+s)^2][1 - r_c^2 \exp(-2(k+s)d)]} \frac{M}{D}, \tag{68}$$

where

$$M = A_1 [A_3 + A_2 r_c - r_d (A_2 + A_3 r_c)] +$$

$$+ [A_1 (A_2 r_d - A_3) \cosh(A_1 d) + (A_2 (A_5 - A_4 r_d) +$$

$$+ A_3 (A_5 r_d - A_4)) \sinh(A_1 d)] \exp[(k+s)d] +$$

$$+ r_c [A_1 (A_3 r_d - A_2) \cosh(A_1 d) + (A_3 (A_5 - A_4 r_d) +$$

$$+ A_2 (A_5 r_d - A_4)) \sinh(A_1 d)] \exp[-(k+s)d] \tag{69}$$

with the other parameters as given above. If the particles are transparent ($k=0$) the solution cannot be obtained directly from the above results, but modified expressions convenient for calculations were given by Maheu et.al.[112] The coefficients s, k and ρ_c can be directly obtained from Lorenz-Mie calculations for a single particle and are the fundamental material parameters of the composite. However, it should be emphasized that the use of Lorenz-Mie theory for this purpose presupposes a dilute mixture of particles in a matrix. Thus the four-flux theory in its present formulation should be valid for low filling factors of the scattering component only. The extension of multiple scattering theory to high filling factors is a complex problem but some progress has been made.[113]

The four-flux theory has been compared[114] to the rigorous multiple scattering calculations of van de Hulst.[20] Good agreement was found if physically realistic values of the parameters η and ρ_d were used. Thus the four flux theory is sufficiently accurate so that it can be used for detailed comparisons with spectrophotometric integrating sphere measurements. Further simplifications of the theory are possible[112] to obtain two-flux theories such as the well-known Kubelka-Munk model.[115] However, two-flux models do not treat the direct and diffuse components separately and thus are not as accurate as the four-flux models.

One can envisage numerous applications for this kind of theory. In particular it can be used to predict the optical properties of heterogeneous materials consisting of particles dispersed in a matrix. Some examples of this kind of material are paint coatings,[116-118] pigmented polymers,[119] Christiansen filters,[120-122] and fibrous materials.[123] Such materials are of interest for improving energy efficiency in certain cases. For example, paints are applied as selective and non-selective absorbers of solar radiation,[9,118] pigmented polymers are of interest in radiative cooling,[10,119] refractive index matching may be used in order to obtain thermochromism[120] and voltage-induced switching between clear and diffusely scattering states[121,122] in windows, and fibrous materials are commonly used for thermal insulation.[123]

In Fig. 15 we show calculations of total transmittance and reflectance for a composite consisting of TiO_2 spheres of diameter 0.23 μm dispersed in polyethylene.

We used the four-flux theory with $\eta=1$ and $\rho_d=\rho_c$. The parameters s, k and ρ_c were obtained from a Lorenz-Mie calculation which used published optical constants[34] for TiO_2 and measured, as well as published,[124] absorption coefficients of polyethylene. The refractive index of polyethylene was taken to be 1.51. The composite has a very high reflectance in the visible and near infrared wavelength range and a high transmittance further out in the infrared, particularly around 10 μm. This makes the foil suitable for radiative cooling purposes. The absorption of solar radiation is low and the high transmittance in the atmospheric window region around 10 μm gives rise to radiative cooling of an underlying emissive material. Calculations for a TiO_2-polyethylene composite with f=0.065 are in good agreement with experimental results in the visible and near infrared spectral regions, as seen in Fig. 15. Further data are given in the chapter on Materials for Radiative Cooling to Low Temperatures.

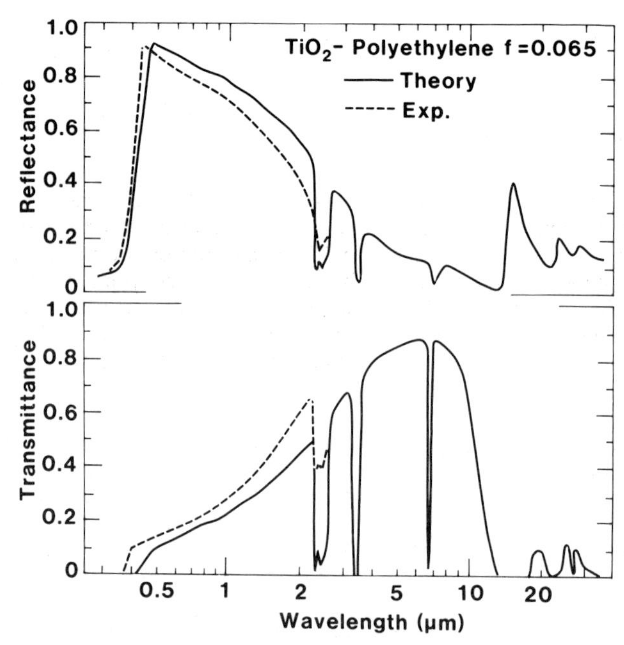

Fig. 15. Total transmittance and reflectance as a function of wavelength for a composite consisting of TiO_2 particles of 0.23 μm median diameter in polyethylene. Dashed curves denote experimental data and continuous curves denote calculations carried out by the four-flux theory with $\eta = 1$ and $\rho_d = \rho_c$. The filling factor of TiO_2 was equal to 0.065 and the thickness of the foil was 100 μm. (From Ref. 119).

Of course, multiple scattering theory has many other applications. These include scattering from geological materials (remote sensing),[22] atmospheric scattering, scattering in seawater, and scattering from biological materials.[21]

VI. CONCLUDING REMARKS

In this review we have considered the theoretical description of the optical properties of inhomogeneous materials. The use of effective medium theories for materials with small inclusions of the components is by now well established. This description is very often used to describe the optical properties of composite films, for example selective solar absorbing coatings. The rigorous Bergman-Milton bounds, on the other hand, are a more recent development and have not yet been so widely used for comparison with experiments. Partly, this is due to an insufficient understanding of the relation between microstructure and specific areas within the Bergman-Milton bounds.

An important challenge for future research is to classify the area within the Bergman-Milton bounds in terms of specific aspects of the microstructures. The first attempts to do this employed calculations of the structural parameter x, which includes contributions from pair and three-point correlation effects, for different model microstructures. The formulation of new effective medium theories taking at least these effects into account appears to be necessary in order to obtain a detailed understanding of optical data.

The optical properties of materials with inhomogeneities with sizes $> \lambda$ have to be described by multiple scattering theory. We treated a simplified four-flux theory which is very useful for comparisons with spectrophotometric measurements. This kind of theory has not been widely used for the description of coatings and optical materials, and further theoretical advances are desirable. In particular, the theory should be extended to the case of dense dispersions of scattering particles. The optical properties of such materials are not well understood at the present time.

Inhomogeneous materials are very important for various applications related to energy efficiency. Many selective solar absorbers consist of very small metal particles in an oxide matrix. The optical properties can generally be accurately described by effective medium models, but in specific cases scattering by rough surfaces is also of major importance. Transparent heat-reflecting metal coatings often have an inhomogeneous network structure which demands effective medium theories or extensions of them. The development of a four-flux theory for materials with large particles is considered to be of great importance for many technological materials. As examples we may cite paints, pigmented polymers and fibrous insulation materials, which all have important applications in energy technology.

Acknowledgement: This work was financially supported by grants from the National Energy Administration of Sweden, the Swedish Natural Science Research Council and the National Swedish Board of Technical Development. Part of the work reviewed in this paper was performed with Prof. C.G. Granqvist, who is thanked for a fruitful collaboration and valuable discussions.

REFERENCES

1. For a recent review see, for example, W.F. Bogaerts and C.M. Lampert, J. Mater. Sci. 18, 2847 (1983). A bibliography was published by G.A. Niklasson and C.G. Granqvist, J. Mater. Sci. 18, 3475 (1983).
2. R.B. Pettit, J.N. Sweet and R.R. Sowell, in *MiCon 82: Optimization of Processing, Properties and Service Performance Through Microstructural Control*, ASTM STP 792, edited by H. Abrams, E. Clark, J. Hood and B. Seth (ASTM, Philadelphia, 1982), p.263; J.N. Sweet, R.B. Pettit and M.B. Chamberlain, Solar Energy Mater. 10, 251 (1984).
3. P.M. Driver and P.G. McCormick, Solar Energy Mater. 6, 159 (1982).
4. G.B. Smith, R.C. McPhedran and G.H. Derrick, Appl. Phys. A 36, 193 (1985).
5. Å. Andersson, O. Hunderi and C.G. Granqvist, J. Appl. Phys. 51, 754 (1980).
6. R.A. Buhrman and H.G. Craighead, in *Solar Materials Science*, edited by L.E. Murr (Academic, New York, 1980), p. 277.
7. G.A. Niklasson and C.G. Granqvist, J. Appl. Phys. 55, 3382 (1984).
8. J.A. Thornton, in *MiCon 82: Optimization of Processing, Properties and Service Performance Through Microstructural Control*, ASTM STP 792, edited by H. Abrams, E. Clark, J. Hood and B. Seth (ASTM, Philadelphia, 1982).
9. S.W. Moore, Proc. Soc. Photo-Opt. Instrum. Engr. 502, 68 (1984).
10. C.G. Granqvist and T.S. Eriksson, this book.
11. G.B. Smith, G.A. Niklasson, J.S.E.M. Svensson and C.G. Granqvist, J. Appl. Phys. 59, 571 (1986).
12. R. Landauer, AIP Conf. Proc. 40, 2 (1978).
13. D.E. Aspnes, Thin Solid Films 89, 249 (1982).
14. G.A. Niklasson and C.G. Granqvist, in *Contribution of Cluster Physics to Materials Science and Technology*, edited by J. Davenas and P.M. Rabette (Martinus Nijhoff, Dordrecht, 1986), p. 539.
15. O. Wiener, Abh. Sächs. Akad. Wiss. Leipzig, Math.-Naturwiss. Kl. 32, 509 (1912).
16. Z. Hashin and S. Shtrikman, J. Appl. Phys. 33, 3125 (1962).
17. D.J. Bergman, Ann. Phys. (N.Y.) 138, 78 (1982).
18. G.W. Milton, J. Appl. Phys. 52, 5286, 5294 (1981).
19. C.F. Bohren, J. Atmos. Sci. 43, 468 (1986).
20. H.C. van de Hulst, *Multiple Light Scattering. Tables, Formulas and Applications* (Academic, New York, 1980), Vols. 1 and 2.
21. A. Ishimaru, *Wave Propagation and Scattering in Random Media* (Academic, New York, 1978), Vols. 1 and 2.
22. W.G. Egan and T.W. Hilgeman, *Optical Properties of Inhomogeneous Materials* (Academic, New York, 1979).
23. G.A. Niklasson, J. Appl. Phys. 62, 258 (1987).
24. F. Lado and S. Torquato, Phys. Rev. B 33, 3370 (1986); S. Torquato and F. Lado, Phys. Rev. B 33, 6428 (1986).
25. J.D. Beasley and S. Torquato, J. Appl. Phys. 60, 3576 (1986).
26. B.U. Felderhof, Physica 126A, 430 (1984).
27. S. Torquato and G. Stell, J. Chem. Phys. 79, 1505 (1983); S. Torquato, J. Chem. Phys. 84, 6345 (1986).
28. A. Hjortsberg, Appl. Opt. 20, 1254 (1981).
29. J.D. Jackson, *Classical Electrodynamics* (Wiley, New York, 1975).

30. M. Born and E. Wolf, *Principles of Optics*, sixth edition, (Pergamon, Oxford, 1980); O.S. Heavens, *Optical Properties of Thin Solid Films* (Butterworth, London, 1955).
31. V. Arkadiev, J. Phys. (U.S.S.R.) 9, 373 (1945).
32. P. Grosse, *Freie Elektronen in Festkörpern* (Springer, Berlin, 1979).
33. D.E. Gray, editor, *American Institute of Physics Handbook* (McGraw Hill, New York, 1963); W.G. Driscoll and W. Vaughan, editors, *Handbook of Optics* (McGraw-Hill, New York, 1978); E.D. Palik, editor, *Handbook of Optical Constants of Solids* (Academic, New York, 1985).
34. J. H. Weaver, C. Krafka, D.W. Lynch and E.E. Koch, *Optical Properties of Metals, Pts. I and II, Physics Data No. 18* (Fachinformationszentrum Energie, Physik, Mathematik, Karlsruhe, 1981).
35. D.J. Bergman, Phys. Rev. Lett. 44, 1285 (1980); Phys. Rev. B 23, 3058 (1981).
36. G.W. Milton, Appl. Phys. Lett. 37, 300 (1980).
37. G.W. Milton and R.C. McPhedran, in *Macroscopic Properties of Disordered Media*, edited by R. Burridge, S. Childress and G. Papanicolaou, Lecture Notes in Physics, Vol. 154 (Springer, Berlin, 1982), p. 183.
38. D.J. Bergman, Phys. Rep. 43, 377 (1978).
39. R.C. McPhedran and G.W. Milton, Appl. Phys. A 26, 207 (1981).
40. G.W. Milton, Appl. Phys. A 26, 125 (1981).
41. K. Golden, J. Mech. Phys. Solids 34, 333 (1986).
42. D.J. Bergman, TAUP 1313-85, unpublished.
43. G.B. Smith, J. Phys. D 10, L39 (1977); Appl. Phys. Lett. 35, 668 (1979).
44. W. Lamb, D.M. Wood and N.W. Ashcroft, Phys. Rev. B 21, 2248 (1980).
45. G.A. Niklasson, C.G. Granqvist and O. Hunderi, Appl. Opt. 20, 26 (1981).
46. C.F. Bohren and D.P. Gilra, J. Colloid Interface Sci. 72, 215 (1979).
47. H.C. van de Hulst, *Light Scattering by Small Particles* (Dover, New York, 1981).
48. R.J. Elliot, J.A. Krumhansl and P.L. Leath, Rev. Mod. Phys. 46, 465 (1974).
49. D. Stroud and F.P. Pan, Phys. Rev. B 17, 1602 (1978).
50. M. Kerker, *The Scattering of Light and Other Electromagnetic Radiation* (Academic, New York, 1969); C.F. Bohren and D.R. Huffman, *Absorption and Scattering of Light by Small Particles* (Wiley, New York, 1983).
51. J.C. Maxwell Garnett, Philos. Trans. R. Soc. (London) 203, 385 (1904); 205, 237 (1906).
52. W.T. Perrins, R.C. McPhedran and D.R. McKenzie, Thin Solid Films 57, 321 (1979).
53. R.C. McPhedran and D.R. McKenzie, Proc. R. Soc. (London) 359, 45 (1978).
54. W.T. Doyle, J. Appl. Phys. 49, 795 (1978).
55. D.R. McKenzie, R.C. McPhedran and G.H. Derrick, Proc. R. Soc. (London) 362, 211 (1978).
56. D.A.G. Bruggeman, Ann. Phys. (Leipzig) 24, 636 (1935).
57. P. Sheng, Phys. Rev. Lett. 45, 60 (1980).
58. T. Hanai, Kolloid Z. 171, 23 (1960).
59. C. Grosse, J. Chim. Phys. 76, 153 (1979); C. Grosse and J.-L. Greffe, J. Chim. Phys. 76, 305 (1979).
60. C. Boned and J. Peyrelasse, J. Phys. D 16, 1777 (1983); Colloid Polym. Sci. 261, 600 (1983).
61. U.J. Gibson, H.G. Craighead and R.A. Buhrman, Phys. Rev. B 25, 1449 (1982).
62. U.J. Gibson and R.A. Buhrman, Phys. Rev. B 27, 5046 (1983).

63. M. Okuyama, K. Furusawa and Y. Hamakawa, Solar Energy 22, 479 (1979).
64. J. Perrin, B. Despax and E. Kay, Phys. Rev. B 32, 719 (1985); E. Kay, Z. Phys. D 3, 251 (1986).
65. L. Martinu, Solar Energy Mater. 15, 21 (1987).
66. R.W. Cohen, G.D. Cody, M.D. Coutts and B. Abeles, Phys. Rev. B 8, 3689 (1973); E.B. Priestley, B. Abeles and R.W. Cohen, Phys. Rev. B 12, 2121 (1975).
67. J.C.C. Fan and P.M. Zavracky, Appl. Phys. Lett. 29, 478 (1976).
68. J.I. Gittleman, B. Abeles, P. Zanzucchi and Y. Arie, Thin Solid Films 45, 9 (1977).
69. C.G. Granqvist, J. Appl. Phys. 50, 2916 (1979).
70. S. Berthier, K. Driss-Khodja and J. Lafait, J. Phys. (Paris) 48, 601 (1987).
71. S. Berthier, K. Driss-Khodja and J. Lafait, Europhys. Lett. 4, 1415 (1987).
72. P. Apell, O. Hunderi and R. Monreal, Phys. Scripta 34, 348 (1986).
73. S. Torquato, J. Stat. Phys. 45, 843 (1986); Phys. Rev. B 35, 5385 (1987).
74. S. Torquato and G. Stell, J. Chem. Phys. 77, 2071 (1982); 78, 3262 (1983).
75. M. Beran, Nuovo Cimento 38, 771 (1965).
76. B.U. Felderhof, J. Phys. C 15, 3943, 3953 (1982).
77. B.B. Mandelbrot, *The Fractal Geometry of Nature* (Freeman, New York, 1983).
78. J.E. Martin, J. Phys. A 18, L207 (1985).
79. A.J. Katz and A.H. Thompson, Phys. Rev. Lett. 54, 1325 (1985).
80. M.V. Berry and I. Percival, Opt. Acta 33, 577 (1986); PM, Hui and D. Stroud, Phys. Rev. B 33, 2163 (1986).
81. S. Torquato and G. Stell, Lett. Appl. Eng. Sci. 23, 375, 385 (1985).
82. S. Torquato, J. Chem. Phys. 83, 4776 (1985).
83. M.N. Miller. J. Math. Phys. 10, 1988 (1969); 12, 1057 (1971).
84. Y. Kantor and D.J. Bergman, J. Phys. C 15, 2033 (1982).
85. A. Liebsch and B.N.J. Persson, J. Phys. C 16, 5375 (1983); A. Liebsch and P.V. Gonzalez, Phys. Rev. B 29, 6907 (1984).
86. M. Gomez and L. Fonesca, Thin Solid Films 125, 243 (1985); M. Gomez, L. Fonesca, G. Rodriguez, A. Velazquez and L. Cruz, Phys. Rev. B 32, 3429 (1985).
87. R.G. Barrera, G. Monsivais and W.L. Mochan, Phys. Rev.B 38, 5371 (1988).
88. V.A. Davis and L. Schwartz, Phys. Rev. B 31, 5155 (1985); Phys. Rev. B 33, 6627 (1986).
89. B.U. Felderhof and R.B. Jones, Z. Phys. B. 62, 43, 215, 225, 231 (1986).
90. S. Torquato, J. Appl. Phys. 58, 3790 (1985).
91. G.A. Niklasson, Solar Energy Mater. 17, 217 (1988).
92. C.G. Granqvist and R.A. Buhrman, J. Appl. Phys. 47, 2200 (1976).
93. S.R. Forrest and T.A. Witten, Jr., J. Phys. A 12, L109 (1979).
94. T. Farestam and G.A. Niklasson, J. Phys. Cond. Matter 1, 2451 (1989).
95. W.T. Doyle, Phys. Rev. 111, 1067 (1958).
96. J. Euler, Z. Phys. 137, 318 (1954).
97. A.G. Mathewson and H.P. Myers, Phys. Scripta 4, 291 (1971).
98. H.G. Hagemann, W. Gudat and C. Kunz, DESY Report SR-74/7 (1974).
99. T.S. Eriksson, A. Hjortsberg, G.A. Niklasson and C.G. Granqvist, Appl. Opt. 20, 2742 (1981).
100. J.T. Cox, G. Hass and J.B. Ramsey, J. Phys. (Paris) 25, 250 (1964).
101. C.G. Granqvist, R.A. Buhrman, J. Wyns and A.J. Sievers, Phys. Rev. Lett. 37, 625 (1976).

102. G.A. Niklasson, S. Yatsuya and C.G. Granqvist, Solid State Commun. 59, 579 (1986).
103. C.G. Granqvist, Z. Phys. B 30, 29 (1978).
104. W.A. Curtin and N.W. Ashcroft, Phys. Rev. B 31, 3287 (1985).
105. A. Wachniewski and H.B. McClung, Phys. Rev. B 33, 8053 (1986).
106. L. Genzel and U. Kreibig, Z. Phys. B 37, 93 (1980).
107. P. Chylek and V. Srivastava, Phys. Rev. B 27, 5098 (1983).
108. D.E. Aspnes, Phys. Rev. B 25, 1358 (1982).
109. W.G. Egan and D.E. Aspnes, Phys. Rev. B 26, 5313 (1982).
110. J.K. Beasley, J.T. Atkins and F.W. Billmeyer, Jr., in *Electromagnetic Scattering*, edited by R.L. Powell and R.S. Stein (Gordon and Breach, New York, 1967), p. 765.
111. P.S. Mudgett and L.W. Richards, Appl. Opt. 10, 1485 (1971).
112. B. Maheu, J.N. Letuolouzan and G. Gousebet, Appl. Opt. 23, 3353 (1984); B. Maheu and G. Gousebet, Appl. Opt. 25, 1122 (1986).
113. G.H. Goedecke, J. Opt. Soc. Am. 67, 1339 (1977).
114. G.A. Niklasson, Appl. Opt. 26, 4034 (1987).
115. P. Kubelka and F. Munk, Z. Techn. Phys. 11a, 464 (1931).
116. L.W. Richards, J. Paint Technol. 42, 276 (1970).
117. T. Kunimoto, H.M. Shafey and T. Teramoto, Bull. JSME 22, 1587 (1979).
118. T. Kunimoto, Y. Tsuboi, S. Iwashita and H.M. Shafey, in *Solar World Congress: Proc. Int. Solar Energy Soc. Congr. Perth 14-19 Aug. 1983*, edited by S.V. Szokolay (Pergamon, Oxford 1984), Vol. 3, p. 1943.
119. G.A. Niklasson and T.S. Eriksson, Proc. Soc. Photo-Opt. Instrum. Engr. 1016, 89 (1989).
120. A.M. Andersson, G.A. Niklasson and C.G. Granqvist, Appl.Opt. 26, 2164 (1987).
121. H.G. Craighead, J. Cheng and S. Hackwood, Appl. Phys. Lett. 40, 22 (1982).
122. F. van Konynenburg, S. Marsland and J. McCoy, Proc. Soc. Photo-Opt. Instrum. Engr. 823, 143 (1987); Solar Energy Mater. 19, 27 (1989).
123. T.W. Tong and C.L. Tien, J. Thermal Insul. 4, 27 (1980).
124. D.R. Smith and E.V. Loewenstein, Appl. Opt. 14, 1335 (1975).

Chapter 3

TRANSPARENT INSULATION MATERIALS

W. Platzer and V. Wittwer

Fraunhofer Institut für Solare Energiesysteme
Oltmannsstr. 22
D-7800 Freiburg, Germany

ABSTRACT

Transparent or translucent insulation materials (TIM's) represent a new class of materials with a high potential for increasing the efficiency of solar thermal conversion systems. A large number of materials have been subjected to theoretical and experimental investigation, as discussed in this chapter. The applications of TIM's include windows, industrial glazings, high temperature collectors, integrated storage collectors, seasonal heat storage systems, and transparently insulated buildings.

I. INTRODUCTION

If one analyses the energy consumption in industrialized countries, one finds that about 40 % of the primary energy is used to produce low temperature thermal energy for heating in dwellings and industrial buildings, domestic hot water and low temperature industrial process heat. This indicates the great potential for thermal solar energy use, and utilisation should be possible even in the Central European climate.

Thermal solar energy is used in passive systems such as greenhouses or windows. In the active systems field, developments have resulted in reliable collector systems for swimming pools and domestic hot water. The most recent work in both areas is on highly efficient collector systems for process heat and new systems like transparently insulated walls which are discussed later.

The fundamental physical principle which is used in all of these applications is the wavelength difference between the solar radiation absorbed by a dark surface, which may be an absorber or aperture to a room, and the thermal radiation which is emitted by the heated absorber or room. This difference allows the utilisation of selective cover layers made of so-called transparent insulation materials (TIM's) which are transparent for the solar radiation and opaque for the thermal radiation. As a result, the IR-radiation losses of the heated system are reduced and the system is more effective than one without a cover. Of course there are other heat loss mechanisms, but all of them are less important. The most prominent candidate for such a selective cover is a glass pane, which fulfils the physical requirements quite well. Therefore a glass pane may be regarded as the simplest type of transparent insulation material. As shown later, a large number of different TIM's are possible, varying in their physical characteristic data and suitable for different applications.[1] From the physical point of view, high transmittance of solar radiation and low thermal conductivity would be the best combination, but other aspects are also important. High transmittance can cause an overheating problem in greenhouses for example. Scattered light might be better than direct light for daylighting applications. Economic aspects may be another reason to use different materials in systems for different applications.

If one looks at the energy efficiency of a thermal system there are two critical characteristics: the function of the solar input which can be absorbed by the system - depending on the solar transmittance of the cover and the absorptance of the absorber - and the portion of the heat produced in the system which can be stored and used - depending mainly on the thermal insulation of the system (U-value). For large energy gains one always needs a high irradiation level and a high $\tau\alpha$ - product, where τ denotes the transmittance of the collector cover and α the absorptance of the absorber plate, but the influence of the U-value is strongly dependent on the temperature level of the system in comparison to the surroundings and the storage time. In systems one therefore has to optimize the TIM for different applications.

The principle characteristics of the new type of TIM will be demonstrated in the following example. A scientist has a black absorber insulated on the backside, an air gap of 10 cm in front of the absorber and a certain amount of glass sufficent to cast a 5 mm thick pane. His aim is to optimise the collector for different applications by varying the spatial distribution of this amount of glass as shown in Fig. 1. Probably he will start with the single glazed collector, will go on to double and triple glazing, but then he will stop for a moment, because he finds that on the one hand the thermal insulation gets better and better but on the other hand the transmittance gets worse. Even if he used expensive antireflection coatings or low-refractive-index materials, at some point there would be a limit in the feasible number of layers for such a system.

To overcome these reflection problems, the scientist has to change the geometry. By rotating the glass by 90°, he no longer has reflection losses, and for non-scattering and non-absorbing materials the transmittance will be close to 1.0. However, he has to do this very carefully. If his structure is open, he will get losses due to convection and, what might be even more detrimental, there will be only low damping of the thermal radiation. By contrast, if he uses honeycomb structures or capillaries with a large aspect ratio, he will get a cover with a high transmittance and a low thermal conductance. As a result, he will measure that now the thermal conductivity in the air, of the material (glass), and the thermal radiation losses will be of the same order of magnitude.[2-4]

Are there any further possibilities to improve the characteristics of the material? From the theoretical point of view, a homogeneous distribution of the 5 mm glass across the total volume of 10 cm should be the best. Indeed there exists such a material with nearly ideal characteristics; it is called an aerogel. The material is produced by a special drying process from sol-gel glass. It is an open porous structure with typically 5 % volume occupied by glass. The pore diameters are well below the wavelengths of visible light, and therefore one finds only a small amount of Rayleigh scattering at short wavelengths. The thermal conductivity of the gas is reduced as the mean free path of the molecules is larger than the pore size, and even the solid state conductivity is diminished by the structure of the material in comparison to glass. In Fig. 1 the different materials are shown schematically and the main heat loss mechanisms are indicated.

Table 1 shows the expected characteristic data for pure glass as the base material and illustrates the principal potential of optimizing a TIM by varying only the spatial distribution of one material. Clearly the heat insulation can be improved by a factor ~40 without changing the transmittance very much.

In practical applications, not only the efficiency but also the simplicity of construction, the long term stability, and the cost are very important. Besides glass, many plastic materials can be used in a similar way and, as shown later, in most real systems a combination with spectrally selective surfaces, air gaps, different gas fillings or even vacuum are used.

Table 1. Ideal values of transmittance and thermal insulation for an arrangement involving pure glass.

Configuration	1 pane	2 panes	3 panes	honeycomb	aerogel
Diffuse transmittance, τ_{dif}:	0.85	0.77	0.71	0.95	0.77
U-value: (W/m^2K)	8	4	2.7	0.8	0.2

II. GENERIC TYPES OF TRANSPARENT INSULATION MATERIALS

Although there is a multitude of very different material types, a classification shall be attempted below. Each class or type then should be described, at least approximately, by a theoretical model. There are many criteria which could be used for a classification such as production process, base materials, environmental hazard associated with the material use and production, fire resistance or mechanical stability. These criteria are certainly important for the application, but for the basic physics they are secondary. A physical criterion could be the thermal emittance of the material, which means a division into low-emitting and high-emitting materials. This will be an important point when the combination of the materials with spectrally selective absorbers will be discussed. With respect to optical qualities one may also classify the materials according to the optical path of

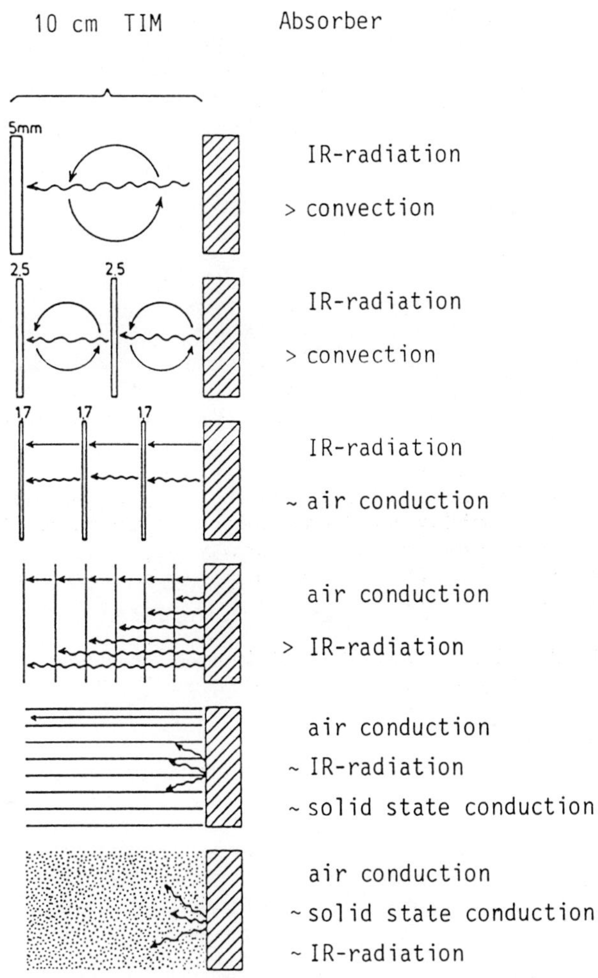

Fig. 1. Sequential refinement of structures made from the same amount of glass, and main heat loss mechanisms.

an incoming beam: Clear glasses for instance can be seen through, but diffusing structures can not. Also there are structures which preserve the incidence angle but still do not allow a clear view. Both criteria are physically meaningful, but there is a continum of materials between the extreme cases, as is made obvious by the emittance. However, mixed types also exist with respect to their optical behaviour. A honeycomb structure produced from weakly scattering plastic foils does not fully preserve the incidence angle for the transmitted beam, but also has a small diffusing component. A classification according to the geometric structure of the materials has proven to be rather useful. Four generic types are proposed, which show different physical behaviour and include most real materials. They are shown in Fig. 2.

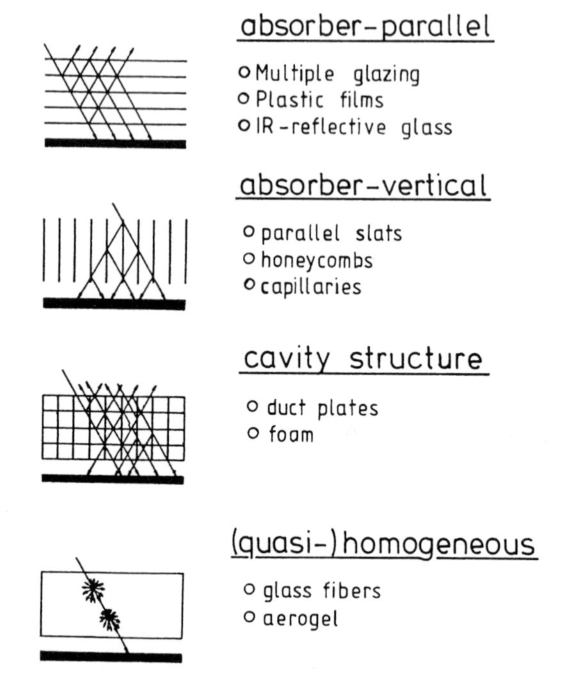

Fig. 2. Generic types of transparent insulation materials.

The first type is the well-known *absorber-parallel* cover with multiple glazings or plastic sheets, which may be clear but also diffusing. High optical reflection losses prohibit the use of a large number of layers. The glass panes or plastic sheets have defined temperatures (approximately constant) but because of convection in the intervening gaps, a one-dimensional temperature distribution cannot be given.

The *absorber-vertical* structures include honeycomb and capillary materials with different cross-sectional geometries as well as slit structures (plastic sheets stretched parallel across the collector). As the incoming beam is reflected and transmitted by the structure walls towards the absorber, optical losses may be very small, and only some scattering and absorption within the sheets reduce the overall transmittance. For clear sheets with low extinction, the transmission properties are nearly independent of the material thickness. Therefore very thick samples may be used. Contrary to the first type, the absorber-parallel structures, where convection is very often present in practice, convection may be effectively suppressed when the aspect ratio (cell length divided by cell width) is chosen well. If both types are combined, one gets a *cavity structure,* represented in reality by transparent multiple duct plates or transparent foam with bubble sizes of some millimetres. From the optical viewpoint, these materials have approximately the same transmittance as an equivalent multiple sheet cover, as the reflection is the dominant loss, but they have the advantage of effectively suppressing the convection.

Quasi-homogeneous layers are characterised by properties similar to those mentioned above, but they stem from other physical mechanisms, namely scattering and absorption. Aerogel, a microporous "silicate foam", belongs to this class. Because of pores with sizes of some 100 Å, light is scattered within the material in analogy with the well-known Rayleigh scattering of the blue sky.[5,6] Glass fibre materials do not have this homogeneity, as single fibres can be seen, but they can be treated and analysed with similar methods.

For each of the four generic types, theoretical approaches exist which are capable of a satisfactory description of the basic features of the materials. Of course transition materials exist which cannot be strictly classified. Folded or V-corrugated foils are such examples. If the corrugation angle is small, the cover behaves essentially like an absorber-vertical material; if the angle is large, it behaves nearly like an absorber-parallel cover. Also honeycomb structures with cells not vertically oriented with respect to the absorber are produced, which provide a transition between the absorber-parallel and the cavity structure type. Nevertheless, the generic types provide a useful approximation of real materials and an understanding of these types is necessary before the more complex materials may be tackled.

III. THEORETICAL MODELS

The aim of a theoretical discription of a TIM is to give possibilities to optimise the physical behaviour of the materials. The choice of base material, the geometrical structure of the TIM and fundamental physical boundary conditions influence heat transport processes and solar transmittance. Therefore these parameters have to be included in the models. However, models are bound to idealise and to describe physical reality only to a certain level of sophistication. Theoretical calculations have the advantage of yielding continuous functions for important parameters, whereas experiments can only give single data points because the number of experiments is limited.

For TIM's certainly the heat resistance R or its inverse, the heat conductance $\Lambda = 1/R$, and the solar transmittance τ are the most important physical values. If systems as a whole are of interest, then the heat transfer coefficients, h_t, between surfaces and air also have to be considered since they influence the heat losses (U-values) and the solar gains (characterised by the total energy transmittance g of the system, which is also called the effective transmittance absorptance product $(\tau\alpha)_e$ in the solar collector field (Ref. 7)). Absorbed solar radiation within the material may flow to different extents to the ambient and to the collector, depending for instance on the emittance of the absorber or the wind velocity at the outer surface. As there are several classes of materials, accordingly different models exist for calculating heat transfer and transmittance.

The first system to be considered is a multiple cover (e.g., double glazing). If the transmittance and the reflectance values for the front and back sides of each single layer are given, then several methods allow the overall properties, including the absorption, for multiple layers to be calculated. We can mention the net flux method of Siegel and Howell,[8] the embedding method of Edwards[9] and a matrix method of Rubin.[10] Of course, the equations of these computational methods are valid only for specified values of incidence angle ϕ, polarisation, and wavelength. However, approximations may be tried for integrated quantities if the accuracy does not need to be too high and the number of layers is small (≤ 3). For dielectric layers, the properties may be calculated easily from the index of refraction and the absorption coefficient (e.g., Ref. 11). If the layers are not clear, but at least partly diffusive, the input for the multiple cover model needs the bidirectional entities $\tau(\phi_{in}, \phi_{out})$, etc., and a matrix calculation has to be used.[12]

Radiation heat transport is easily calculated within a multiple glazing by a formula of Pflüger.[13] The heat transfer, including the coupled convective and infrared radiation transport, was first described by Hollands and Wright.[14] If the absorption within the different layers is known, the g-value may also be calculated.

The methods for multiple cover materials are developed, and the choice of method and degree of sophistication depend on the question considered.

For thicker materials, where an appreciable temperature distribution within the layer may be noticed, other models should be used. A relatively easy method exists for "grey" absorbing and scattering layers (like aerogel and fibre insulation), when the spectral dependence can be ignored. The IR-radiation transport may be accurately described by methods developed for heat transport in gases and stellar atmospheres.[8,15] Isotropic scattering is easy to handle even with analytical methods. However, if the phase-function of the scattering particle is anisotropic, either an approximate rescaling may be tried or one must resort to numerical or Monte Carlo methods.[16,17]

However the spectral variations in extinction, which is the sum of absorption and scattering, may cause problems. The question of the "true" equivalent extinction coefficient for "grey" models is still under discussion.[16,18]

When the layer is not a homogeneous material, but a geometrical structure with vertical walls like a honeycomb, the angular variation of emission and transmission within the structure is important. The IR-radiation transport can be calculated very well from a model similar to the one used for solar transmittance,[17,19,20] where the angular variation of emittance is explicitly taken into account.[21] The IR-transmittance is needed as input for the reformulated radiation transport equation of a thick layer.[22] The coupling between the conduction within sheet and air and the radiation transport can be solved by an analytical method developed by Cess[23] for homogeneous media, where an effective volumetric absorption coefficient has been defined.[17]

The agreement between experimental data and theoretical calculations is remarkably good, as shown in Fig. 3. If solar radiation is present, volumetric heating of the materials by absorption also has to be considered. This can be done, at least approximately, with the methods mentioned, so the secondary heat gain from solar absorption also may be calculated.

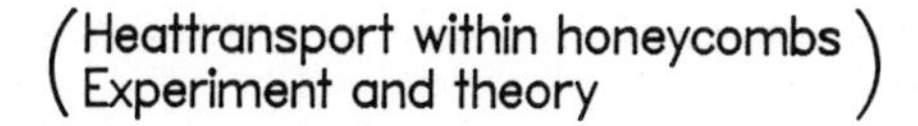

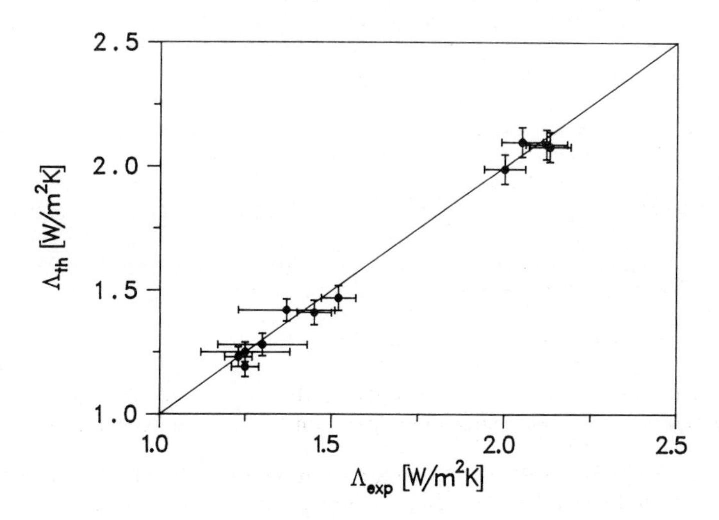

Fig. 3. Comparison between experimental and theoretical values of the heat conductance for different honeycomb materials with aspect ratios between 5 and 22.

A problem for most calculations is the choice of proper input data. The geometric structure of real materials is very often far from regular, and in particular honeycombs may have varying sheet thicknesses and aspect ratios. Also, getting reliable IR-absorption data for materials like aerogel, which can have very different spectral bands, is difficult. Because of these problems, it is not yet clear when non-spectral calculations with averaged input data are sufficient and when exact spectral calculations are necessary. Results for honeycombs and aerogels indicate that the first material type is rather well described by the simpler method, whereas non-spectral methods often produce large errors for aerogels.[17,18]

IV. EXPERIMENTAL CHARACTERIZATION

Several basic experimental set-ups for measuring the fundamental physical values are described in the following section. Other experiments, yielding spectral information, scattering phase functions etc., are sometimes necessary to fully understand a material like aerogel for example,[24] but a discussion of these aspects would go beyond the scope of this chapter.

A. Measurement of Solar Transmittance and Reflectance

For solar energy applications, it is not necessary to be able to see clearly through a collector cover. Instead, all of the transmitted light reaching the absorber is of relevance, even when this light has been scattered or reflected. Therefore the direct-hemispherical or diffuse transmittance $\tau_{dif}(\phi, \rho)$, depending on the incidence angles ϕ (polar) and ρ (azimuth), is important. Very often an azimuthal dependence does not exist (e.g., for usual glazings) or is negligible (for most honeycombs).

Diffuse transmittance can be measured by an integrating sphere, for which the light entering from all directions contributes to a homogeneous radiance of the sphere walls (apart from certain spots) because of multiple reflections.[25-27] Small spheres for spectrometers commonly use a double-beam geometry, whereas large spheres suitable for TIM's are single-beam devices, as shown in Fig. 4. Here a detector measures the radiance of a selected part of the sphere wall, where no direct light from the entrance port should be incident. Measurements with and without the sample in front of the entrance port give a raw value for the transmittance, which has to be corrected because the sensitivity of the sphere changes somewhat depending on the reflectance of the sample. A collimated beam of diameter d irradiates a sample with dimension D, having a structure with typical dimension δ (say the bubble diameter of an acrylic foam or the cell width of a honeycomb). The radiation traverses the sample and hits the sphere entrance port with diameter a. Now two different set-ups are conceivable:[17,20,28,29] Either the beam diameter is much smaller than the entrance port and the sample ($D \approx a >> d >> \delta$), *or* a large sample is homogeneously irradiated ($D \approx d >> a >> \delta$). As the

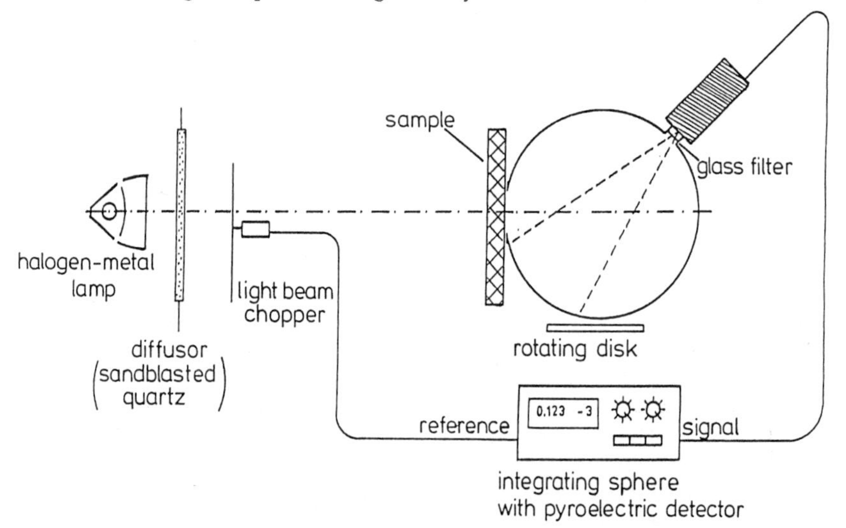

Fig. 4. Experimental set-up for an angle-dependent transmittance measurement using an integrating sphere. The configuration uses a single-beam and large area illumination.

sphere diameter has to be much larger than the diameter of the entrance port, to keep the corrections small, the first set-up calls for very large spheres, whereas for the second set-up the requirements on radiation homogeneity and intensity may be difficult to meet. It is important for both designs to have a strong radiation source for parallel light (lamp with collimator), with a spectral intensity distribution close to that of the terrestrial solar spectrum, and a spectrally non-selective detector for the region 0.3 to 2.5 μm. Pyroelectric detectors offer the best sensitivity, but thermopiles may be also used if the power of the illumination is large enough. The coating of the integrating sphere should be a diffusing one with high reflectance in the whole relevant wavelength region. For very large spheres, inner coatings of Halon (pressed Polytetrafluorethylene powder, PTFE) is the best choice,[30] but is difficult to apply. For the same reason, smoked MgO is preferable only for small spheres. $BaSO_4$-powder-based paints are relatively easy to apply, but a binder other than water should be used because of the strong absorption bands of water in the near infrared.

Reflectance measurements in principle can also be made with an integrating sphere, with an additional sample port in the rear part of the sphere. The light enters the front entrance port and is reflected from a reference surface or from the sample into the sphere. For each incidence angle one needs a separate opening for the incoming light. A further problem stems from the scattered light in thick samples: radiation, although scattered or reflected back towards the sphere, may not return through the sample port. Large sample ports and small beam diameters are usually necessary, so that the lost part of the radiation is not too large. As a consequence the measured reflectance for thick scattering samples is often smaller than the true value.[31]

B. Measurement of Heat Transport

For the measurement of total heat transport within TIM's, in principle any hot-plate apparatus for determination of U-values can be used, as described in the standards (ISO, DIN 4757 (German Industrial Standard)). Since the materials are partially transparent to IR-radiation, the emissivity of the surfaces in the apparatus should be known and be interchangeable.

To determine convection properties of the samples for varying inclination, the whole apparatus should be rotatable by 180 degrees. The 180° position (hot plate on top of a horizontal material) is used to determine the heat transport without convection. The difference for any other position to this value gives the convective part of the heat transport and may be expressed in terms of the Nusselt number Nu. For exact and fast measurement, a hot-plate apparatus with heat flux meters on both plates is advantageous (e.g., Ref. 13). Usually results are shown as a function of the dimensionless Rayleigh number Ra, which incorporates thickness and temperature difference across the material.

Figure 5 shows the Nusselt numbers for 10 cm square honeycombs with aspect ratios (thickness divided by cell width) 5 and 10 and inclination 45°. One can see that for the second sample, convection is negligible for solar thermal application (below a temperature difference between the hot and the cold side, ΔT, of 100°C). It should be noted, however, that this observation is valid only for the material thickness 10 cm. For thicker materials, the Ra number increases rapidly, whereas for 5 cm thickness, for example, an aspect ratio of 6 to 8 may be enough to suppress convection. Results show also that the maximum of the Nusselt number for constant Ra is between 30° and 45° inclination angle (Fig. 6).

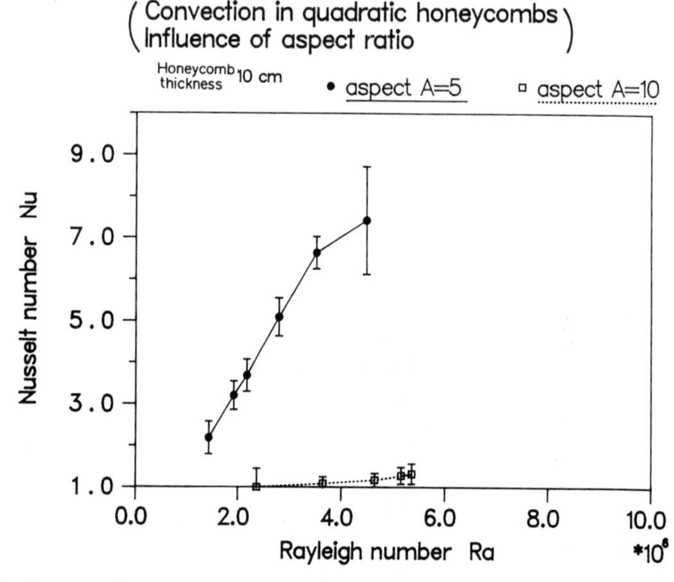

Fig. 5. Convection suppression of square honeycomb materials (10 cm thickness) with aspect ratios of 5 and 10 expressed by Nusselt number (Nu = 1 means no convection).

When the heat transport of a collector type system is measured, convection is present not only within the materials but also in the air gaps. The main impact of introducing air gaps in systems with a low emissive surface is to reduce the radiation-conduction coupling between the absorber surface and TIM.[32,33] Figure 7 shows measured and calculated U-values for a set-up with a 10 cm honeycomb material, a variable air gap, and a spectrally selective absorber having a hemispherical emittance, ε_h, of 0.08. No convection is present in the system. As a comparison, the dashed line shows the hypothetical additive U-value if no coupling existed. For high-emissive plates (i.e., $\varepsilon_h > 0.70$) this is a good approximation.

If convection is present in the air gap, as is usual for inclined collectors, the decoupling effect of the air gap is decreased but still appreciable (Fig. 8).

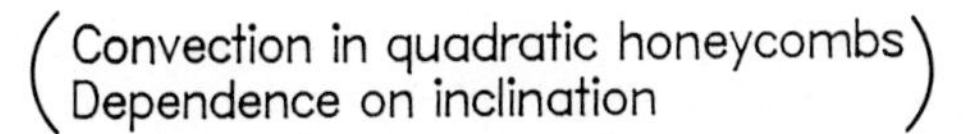

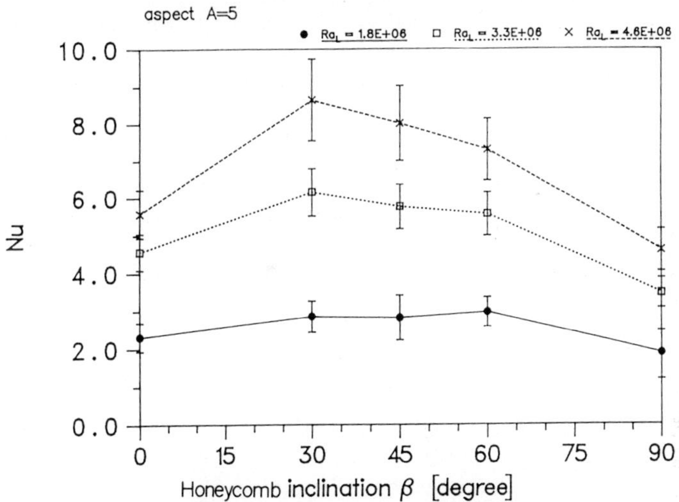

Fig. 6. Dependence of convection within the honeycombs on the inclination of the absorber system (aspect ratio 5, material thickness 10 cm).

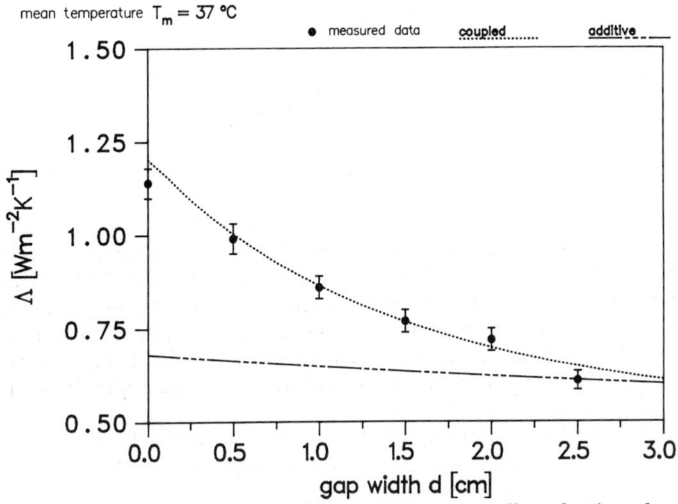

Fig. 7. Heat transmittance coefficient for a spectrally selective absorber system with 10 cm honeycomb and variable air gap (no convection). Dashed lower curve represents a calculation neglecting the radiation-conduction coupling.

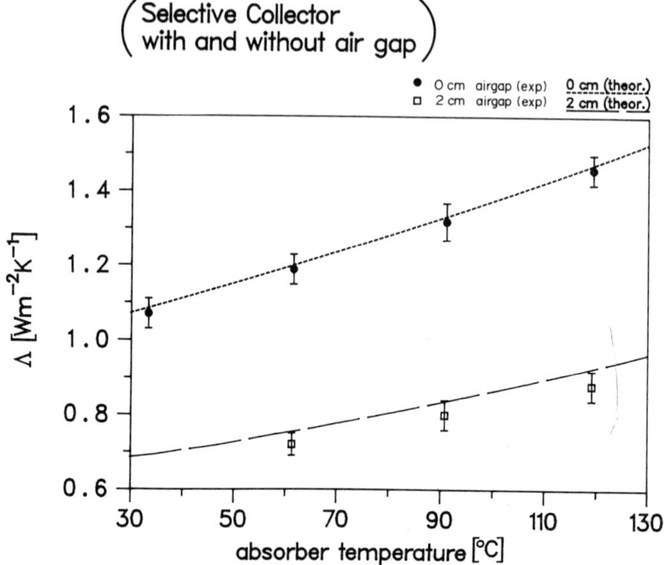

Fig. 8. U-values of spectrally selective honeycomb collectors with and without air gap. The inclination is 45° and convection is present.

C. Total Energy Transmittance

Because of solar absorption in most TIM's, solar irradiance will result in a temperature rise within the materials. This leads to a superposition of a heat flux q_{in} towards the absorber, which is proportional, at least to a first approximation, to the irradiance I. Therefore the total energy transmittance g, defined by

$$g(\phi) = (\tau\alpha)(\phi) + q_{in}/I = (\tau\alpha)_e, \qquad (1)$$

is the relevant parameter for thermal systems. Of course, q_{in} depends on the heat resistances from the TIM to the absorber and the environment, but also on the absorber reflectivity and on temperature levels. Strictly speaking, the g-value is a system parameter, and not a material parameter like the solar transmittance. However, calculations show that the latter system parameters are of minor influence, and corrections for the first ones can be introduced if a measurement of g for defined conditions is done.[17,20] Only a few direct measurement methods, all of which are calorimetric, are known. While two of them employ constructions such as water or air collectors, and the fluid temperature difference between inlet and outlet together with the flow rate are measured, the absorber of the apparatus at the Institute for Solar Energy Systems (ISE) in Freiburg, on the other hand, uses heat flux meters mounted on a cooled copper plate.[34] The advantages of this device lie in a relatively fast response, a good accuracy, and the possibility of measuring with different incidence angles.

In principle, the measurement is analogous to an efficiency determination of a solar collector: The ambient temperature T_a, the absorber temperature T_{abs}, and the irradiation I are needed as well as the net heat gain q_{net} of the absorber. The collector equation

$$q_{net} = gI - U(T_{abs} - T_a) \tag{2}$$

then can be used to determine g, if the U-value of the TIM-cover is known. This can be measured with the same apparatus without irradiation. One advantage of the ISE-device certainly is the direct determination of T_{abs} and of q_{net} with heat flux meters, whereas elsewhere the usual heat removal factor F_R (see Ref. 7) has to be taken into account. The total energy transmittance for diffuse isotropic irradiation, g_{dif}, can also be determined by integrating numerically over the angular dependent $g(\phi)$. Results appropriate for TIM's have become available now; for these g is given under normalised conditions with maximum absorptance (α) equal to unity, ambient temperature (T_a) of 26°C, absorber temperature (T_{abs}) of 30°C, internal heat transfer coefficient (h_i) of 6.5 ± 0.5 W/m^2K, and external heat transrfer coefficient (h_a) of 8 ± 1 W/m^2K.[34] Representative data are given in Table 2.

Still, there is not enough data with the desired accuracy to determine the effective scattering part of the extinction for different base materials, e.g. for the honeycomb design. Only with additional work, and with theoretical models, can the impact of a material change on the total energy transmittance be judged, and therefore the impact on the layout of a solar system. An example for the maximum difference between solar and total energy transmittance is given in Fig. 9, where both properties are plotted as functions of the incidence angle. The difference for diffuse irradiation is an 8 percent change in transmittance.

Table 2 Solar and total energy transmittance for diffuse irradiation of different types of materials.

Material	Thickness (cm)	τ_{dif}	g_{dif}
Floatglass	0.3	0.74 ± 0.03	0.79 ± 0.03
PMMA pane	0.3	0.77 ± 0.03	0.81 ± 0.03
PMMA foam	1.5	0.55 ± 0.03	0.57 ± 0.02
Aerogel granules between PMMA	0.3/2.0/0.3	0.37 ± 0.03	0.42 ± 0.03
PC honeycombs	10.0	0.78 ± 0.03	0.82 ± 0.05
PMMA capillaries	9.8	0.74 ± 0.03	0.77 ± 0.05

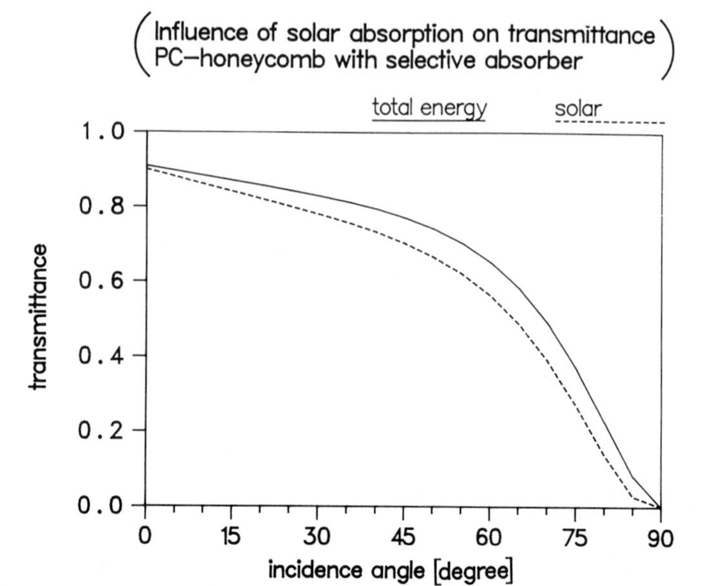

Fig. 9. Comparison of calculated solar and total energy transmittance for 10 cm honeycomb. Absorption, but no scattering in the plastic sheets, is assumed.

V. AVAILABLE MATERIALS AND REAL TRANSPARENT COVER SYSTEMS

A large number of different materials and systems have been tested on a laboratory scale in the last years, but only a few systems are available on the market. In the following sections some of these will be examined in more detail.

A. Multiple Glazing with Low Refractive Index Materials

For flat plate solar collectors working without storage in the medium temperature range ($\Delta T \approx 50°C$), a high g-value and therefore a large solar transmittance is very important, whereas a low U-value is not so meaningful. For these applications, a spectrally selective absorber has been used in most cases, together with a highly transparent insulation material to suppress convection.

Parallel PTFE (Teflon) slats and honeycomb systems were investigated in Australia and Canada, and Swedish companies have installed large area collector fields with different arrangements of Teflon foils in the gap between the absorber and the cover glass. Characteristic data of such systems are shown in Table 3, which illustrates diffuse solar transmittance and U-values for a number of different systems.

In applications, the cost of the systems is an important point. The first field tests with large area collectors in Sweden indicate that for district heating systems with seasonal storage a single Teflon film is the best option at the moment.[35] A comparison between honeycombs and parallel slat structures shows that the results could be very similar for both systems.[36]

Table 3. Characteristic data of collector cover systems with PTFE sheets between absorber and cover glass. The configuration had $\varepsilon_h \approx 0.23$ and $\Delta T \approx 55°C$ (excepting the U-value marked by an asterisk, for which $\varepsilon_h \approx 0.9$).

Sheet type:	Single	Double	V-corrugated	Honeycomb	None
τ_{dif}:	0.82	0.75	0.83	0.81	0.85
U-value: (W/m^2K)	2.9	2.5	2.75	2.0*	3.7

B. Honeycomb and Capillary Structures

For systems which are working at higher temperatures or with internal storage, a low U-value becomes more and more important. Of course a high g-value is always good. For real systems one has to optimize the material again. For these applications, IR-opaque honeycomb and capillary structures were developed; they suppress the convection and IR-radiation very effectively.[37] For low temperatures, even a spectrally selective coating is no longer necessary.

A suppression of IR-radiation is much harder to achieve than a prevention of convection. The materials optimisation process therefore is a question of how large an aspect ratio has to be chosen and how the thickness of the sheets should be adapted to get good damping for the IR-radiation on the one hand and not too high absorption and scattering for the incoming solar radiation on the other hand. In most of these systems, convection in the structure is no longer a problem.

Two types of materials are available in large quantities on the market: a square honeycomb material (3.5×3.5 mm^2 cell size) made from polycarbonate, and capillary materials made from different plastics with various capillary diameters. Typical thicknesses of the plastic layers in both cases is 20 to 50 µm. In principle, the materials can be optimized for a special application. In Table 4 typical characteristic data are given for a number of these materials without an additional cover glass, required for a real application. Thermal data were measured with black absorbers ($\varepsilon_h \approx 0.9$) on both sides. The material had a mean temperature (T_{mean}) of 10°C.

The data show the high potential of these materials for high temperature or storage systems. Further improvements might be possible in the future, mainly in transmittance as a result of better production technologies, but also in U-values which could be lowered by optimisation of the geometry.

Table 4. Characteristic data of honeycomb and capillary structures. The configurations had $\Delta T \approx 10°C$ and $T_{mean} = 10°C$.

Sample:	Honeycomb. 5 cm thick Polycarbonate	Honeycomb. 10 cm thick Polycarbonate	Capillaries. 10 cm thick Polycarbonate	Capillaries. 10 cm thick PMMA
τ_{dif}:	0.85	0.78	0.73	0.80
Λ (W/m^2K):	2	1.07	0.98	0.91

C. Homogeneous Materials (Aerogels)

As stated in the introduction, a homogeneous material distribution would be the optimal one from the physical point of view. This brings us to aerogels, which have been known for more than 50 years. The first applications in the optical field were with experiments in Cerenkov counters in elementary particle physics.

The utilization of these transparent and well insulating materials in window and cover systems started about 10 years ago. Yet still only two different types of aerogels are available on a small scale for tests. One type comes as tiles of typically 2 cm thickness and dimensions of up to 60 x 60 cm, which are highly transparent and can be used in windows. The other type comes as granules of variable diameter (typically 1 to 10 mm), which should be much cheaper in production and which can be filled into the gap between panes in a double glazing, for example. This latter material as yet shows strong scattering but it could be used in collectors and facades for example, as will be discussed later. Table 5 shows characteristic data of a double glazed window with different aerogel fillings. In principle, systems using tiles or granules can be improved by evacuation or special gas fillings.

Table 5. Characteristic data of aerogel samples between panes in a double glazing. The configurations had $\Delta T \approx 10°C$ and $T_{mean} = 10°C$.

Sample:	Tiles	Granules with diameters: 6-8 mm	4-6 mm	3-4 mm	<2 mm
τ_{dif}:	0.57	0.43	0.42	0.40	0.22
Λ (W/m^2K):	0.95	1.15	1.13	1.03	0.98

VI. APPLICATIONS OF TRANSPARENT INSULATION MATERIALS

A. Flat Plate Collectors for Process Heat (90 to 150°C)

Flat plate collectors can be tremendously improved by using transparent insulation materials as the front cover. A reasonable criterion for the comparison of process heat collectors is the efficiency at an irradiation of 800 W/m^2 and a temperature difference of $\Delta T = 100°C$. Under these conditions, a conventional flat plate collector with a spectrally selective absorber and single glazing has an efficiency of approximately 20 % only, whereas vacuum tube collectors reach about 50 %. Using a 10 cm honeycomb structure made of polycarbonate, the front cover losses of a flat plate collector can be reduced to as little as 0.9 W/m^2K (at $\Delta T = 100°C$). Such a collector has been developed recently at ISE.[38] A collector field of 60 m^2 area has been tested in combination with a transmembrane water desalination unit on the Canary Islands since May 1988. Figure 10 shows a comparison of data for flat plate, vacuum tube, and flat plate honeycomb collectors.

According to calculations, still further improvements of optimized flat plate collectors with TIM's are possible, so that they will even outperform vacuum tube collectors in the future.

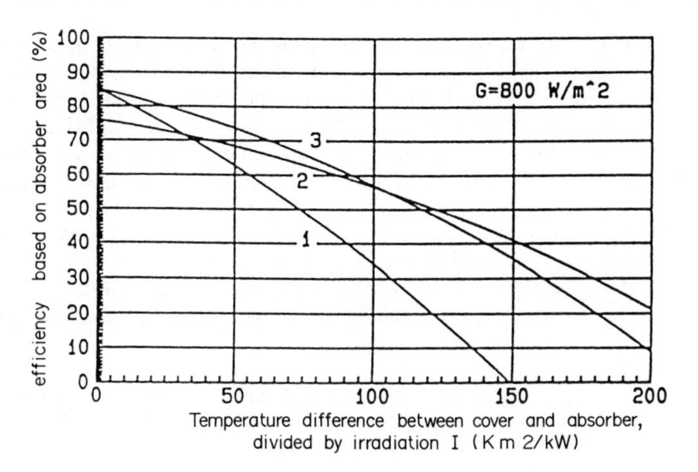

Fig. 10. Efficiency curves for solar collectors of different designs.
1 = spectrally selective flat plate
2 = spectrally selective with honeycomb
3 = vacuum tube

B. Integrated Storage Collectors for Domestic Hot Water

An integrated storage collector (ISC) consists of a flat storage tank whose upper surface acts as an absorber and is supplied with transparent insulation. A typical design is shown in Figure 11.[39] The tank is constructed to withstand water pressure from the mains. The advantages over a conventional solar domestic hot water (DHW) system are

- compact construction, only one single device,
- no heat exchanger, no pump, no control system,
- no anti-freeze fluid.

These advantages can help to bring down the costs of solar domestic hot water systems. Up to now, integrated storage collectors with simple single or double glazing could be used successfully in warm climates only. Under Central European winter weather conditions, freezing occurs if simple glazing is used. These difficulties can be overcome, however, by using transparent insulation materials as the covers of ISC's. Through extensive computer simulations, it has been shown that for ISC's there is no danger of freezing in Central European weather.

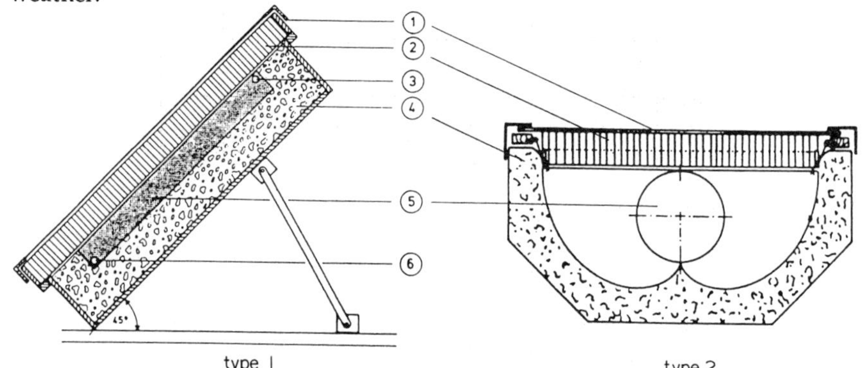

Fig. 11. Cross-sections of two types of integrated storage collectors designed to operate under mains pressure.
1 = glass cover
2 = transparent insulation
3 = water outlet
4 = opaque insulation
5 = water tank
6 = water inlet

Three prototypes of integrated storage collectors with TIM (g = 0.6 and 1.7 W/m^2K overall heat loss coefficient with respect to absorber area) have been constructed at ISE and are currently under investigation. In a computer-controlled test routine, 40 litres/m^2 of hot water are withdrawn every day from the integrated storage collector. The prototypes are characterized in terms of system efficiency and solar fraction. The system efficiency is the yearly accumulated heat gain (i.e., the energy content of tapped water at the temperature level of the storage of the solar system minus the energy content of the cold mains water) divided by the yearly accumulated solar radiation energy on the collector area, and the solar fraction is the yearly accumulated heat gain minus storage losses divided by the yearly accumulated heat demand (i.e., the energy content of tapped warm water delivered by the solar system as well as heated up by an auxiliary heater minus the energy content of the cold mains water). During nine months of operation (from November '86 to July '87) the measured system efficiency was 31 % at a solar fraction of 52 % (Fig. 12). More details of the test results are described by Schmidt et.al.[40] The efficiencies reported in Fig. 12 are considerably higher than those of conventional solar DHW systems with flat plate collectors under Central European weather conditions.

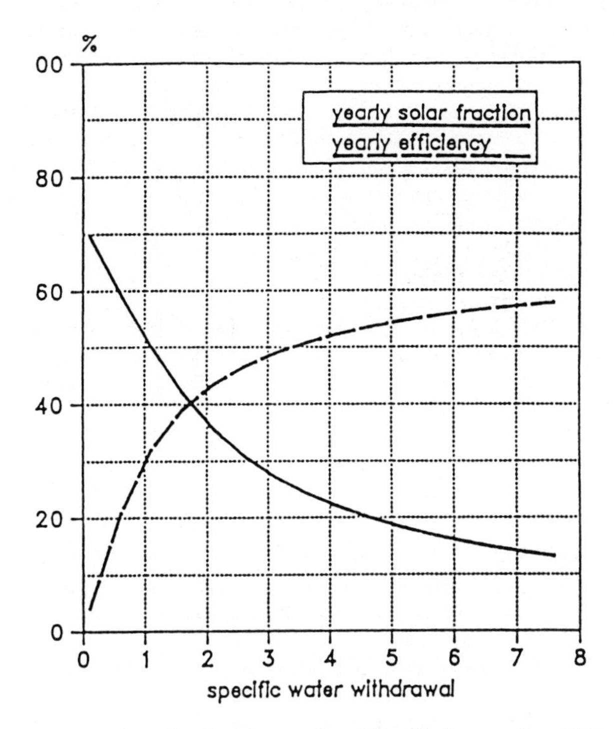

Fig. 12. Yearly solar fraction and yearly efficiency of an ISC (type 2) as a function of the specific water withdrawal.

C House Heating with Transparently Insulated Collector-Storage Walls

A new concept of external transparent insulation of massive house walls combines the advantages of conventional opaque insulation (reduction of heat losses) with those of a solar collector system (conversion of solar radiation into useful heat). It is therefore practical not only in cold and sunny climates, but also in northern climates. Only the special properties of good TIM's allow an application of this principle under the latter conditions.

The working principle is shown in Fig. 13. A layer of transparent insulation is mounted on the facade of a building. The surface of the wall has the properties of an absorber, so the incoming radiation is converted into thermal energy. Depending on the relation of the U-values of the wall and the transparent cover, more or less energy will flow into the building, thus contributing useful heat. A steady-state model was developed by Goetzberger et.al.[41,42] and has been compared with experimental and numerical results.[43,44]

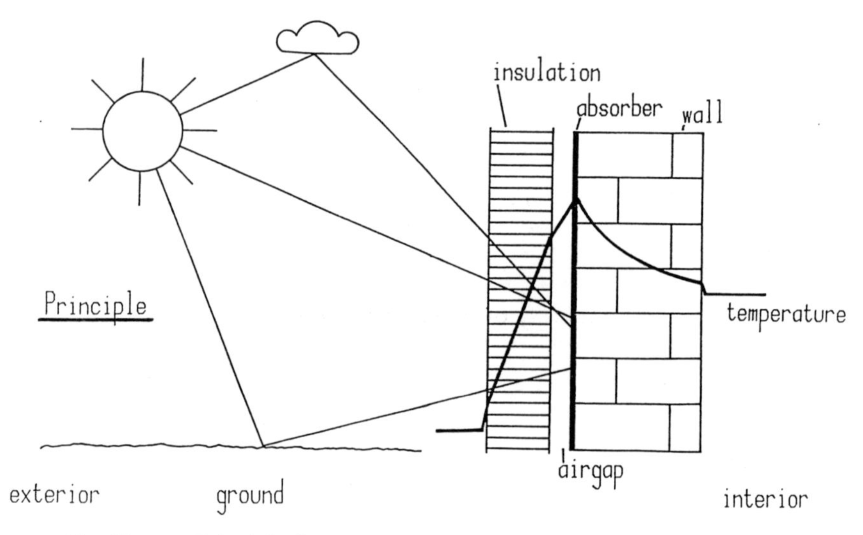

Fig. 13. Principle for transparent insulation of massive house walls.

The theory describes monthly, or even weekly, mean values very well and is straightforward as long as no switching control of the system produces non-linear thermal behaviour. The net heat flux density (q_w) of the system to the outside is given by the total U-value (U_{tot}) and the efficiency (η_0), where U_{tot} accounts for the U-values of the wall construction (U_{wall}) and of the TIM-cover ($U_{TIM\text{-}cover}$) according to

$$U_{tot}^{-1} = U_{wall}^{-1} + U_{TIM\text{-}cover}^{-1}. \qquad (3)$$

The formula for q_w is therefore

$$q_w = U_{tot} (T_i - T_a) - \eta_0 I \qquad (4)$$

with

$$\eta_0 = \frac{g\, U_{wall}}{U_{wall} + U_{TIM\text{-}cover}}, \qquad (5)$$

where I is the solar irradiation intensity as before.

Although the intrinsic storage capacity of the collector wall distributes the solar gains over an extended period of time, producing high comfort conditions, sometimes the excessive solar gains are neither necessary in order to keep the temperature at the required comfort level nor can they be stored within the thermal mass of the building long enough. These parts of the solar gains are then lost again to the surroundings without having reduced the auxiliary heating demand. Under these conditions, the utilization of the solar gains is below one. The utilization factor can only be determined by experiments or numerical simulations. It is normally larger than for direct-gain systems, where very often hot air has to be vented. The useful monthly excess solar gain Q_{sol} (represented by the second part of Eq. 4, i.e., η_0 I) is limited by the monthly remaining heat load L_H of the house. Therefore the ideal linear relation in Eq. (4) changes in reality when Q_{sol} approaches the order of magnitude of L_H, as apparent from Fig. 14. It can also be shown that, for available materials, the system works very well for all orientations except facing north (in the northern hemisphere).

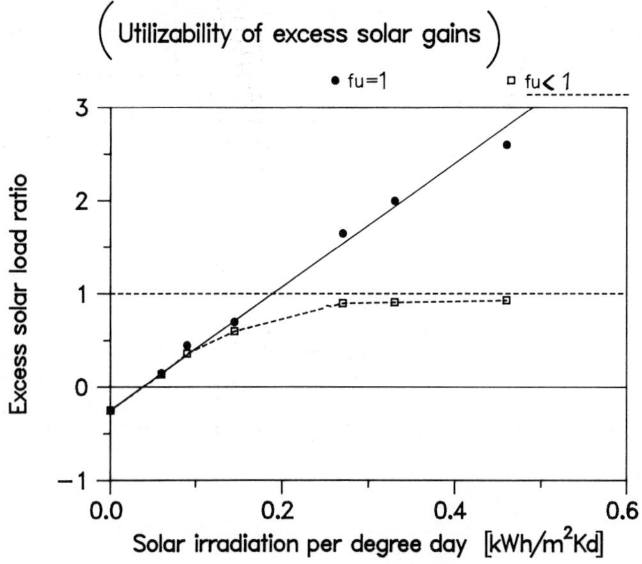

Fig. 14. Excess solar load ratio (integrated positive net heat flow into a building divided by its remaining heat load) vs. irradiation energy divided by degree days (monthly values). Single points: simulation results; continuous line: analytical result using Eq. (4).

Experimental results for a west facade of a house in Freiburg, Germany, have been collected since 1983. Although the house is shaded by neighbouring houses and trees, a comparison of the TIM-system with a conventional opaque insulation and the wall alone shows the remarkable potential for energy savings by use of novel collector-storage walls (Fig. 15).

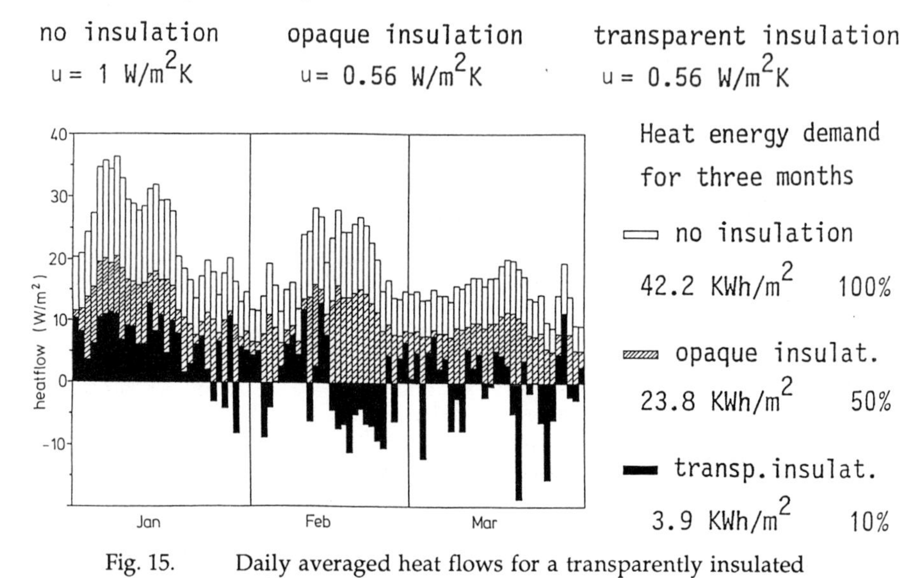

Fig. 15. Daily averaged heat flows for a transparently insulated wall, an equivalent opaque-insulated wall, and a wall without insulation.
(West facade of a house in Freiburg; experimental results for January-March 1985).
Wall: 36 cm brick.
The heat energy demand is the integrated heat flow over 3 months.

One problem, which almost certainly arises, is overheating in the summer. As the collector wall is an efficient system, excess heat will be produced constantly in the summer. Sometimes this problem can be solved by passive shadowing, e.g., with overhangs or by natural venting of excess heat at the absorber, but in most cases an active system is needed. Therefore an automatic roller blind system, controlled by a microcomputer system, was integrated in the test facade. During summer the roller blind is closed. At the beginning of the heating season, the outdoor temperature decreases and the roller blind opens during the day.[40,45]

Since 1988, several buildings of different types have been equipped with TIM-systems and measurements are taken to check the performance of these systems integrated in real houses with inhabitants.

VII. CONCLUSION AND OUTLOOK

The experimental, theoretical and practical work over the last few years has led to a basic understanding of TIM's. The materials now available reach heat transmittance coefficients below 1.0 W/m^2K with transmission values of 0.7. The theoretical models have proven their validity and can be used in simulation and optimization programs. Further improvements, principally in systems but also in materials, can be expected.

The first applications of TIM's, e.g., for flat plate collectors, have shown that an increase in efficiency can be achieved. In other fields, e.g., house heating and storage, completely new concepts are made possible.

One main goal for the future will be to improve the systems and reduce cost, thus enabling the storage of heat from summer to winter.

Initial calculations show the large potential of transparently insulated energy storage systems (Fig. 16). Further improvements of such systems could help to solve the storage problem in large solar heating installations.[46-48]

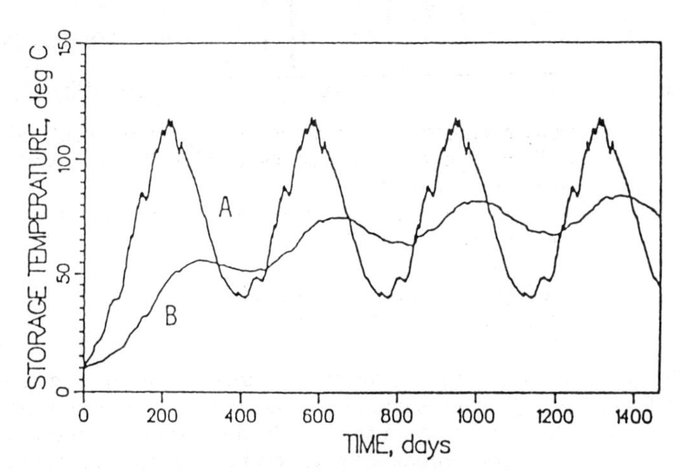

Fig. 16. Simulation results for large transparently insulated storage tanks without user (no water withdrawal).
$A = 200\ m^3$
$B = 30000\ m^3$
Horizontal TIM cover.

REFERENCES

1. V. Wittwer, W. Platzer, A. Pflüger, W. Stahl and A. Goetzberger, in *Proc. ISES Solar World Forum*, Montreal,Vol. 2, p. 1114 (1985).
2. G. Francia, in *Proc. U.N. Conf. on New Sources of Energy*, E/Conf. 35/S/71 (1961).

3. K.G.T. Hollands, Solar Energy 9, 159 (1965).
4. H. Tabor, Solar Energy 12, 549 (1969).
5. A.J. Hunt, Lawrence Berkeley Laboratory Report LBL-15756 (University of California, 1983).
6. M. Rubin and C.M. Lampert, Solar Energy Mater. 7, 393 (1983).
7. J.A. Duffie and W.A. Beckman, *Solar Engineering of Thermal Processes* (Wiley, New York, 1980).
8. R. Siegel and J.R. Howell, *Thermal Radiation Heat Transfer*, second edition (McGraw-Hill, New York, 1981).
9. D.K. Edwards, Solar Energy 12, 401 (1977).
10. M. Rubin, Int. J. Energy Res. 6, 123 (1982).
11. M. Born, *Optik - Ein Lehrbuch*, third edition (Springer, Berlin, 1981).
12. K. Papamichael and F. Winkelmann, Lawrence Berkeley Laboratory Report LBL-20543 (1986).
13. A. Pflüger, Diplomarbeit, University of Freiburg, Germany (1984).
14. K.G.T. Hollands and J.L. Wright, Solar Energy 31, 211 (1983).
15. M.G. Kaganer, Opt. Spektrosk. 26, 443 (1969).
16. R. Caps, Ph.D. Thesis, University of Würzburg, Germany (1985).
17. W. Platzer, Ph.D. Thesis, University of Freiburg, Germany (1988).
18. A. Pflüger, Ph.D. Thesis, University of Freiburg, Germany (1988).
19. K.G.T. Hollands, K.N. Marshall and R.K. Wedel, Solar Energy 21, 231 (1978).
20. W. Platzer, Solar Energy Mater. 16, 275 (1987).
21. M. Rubin, Solar Energy Mater. 6, 375 (1982).
22. D.K. Edwards and R.D. Tobin, J. Heat Transfer (1967), p. 132.
23. R.D. Cess, Adv. Heat Transfer 1, 1 (1964).
24. J. Fricke, editor, *Aerogels*, Springer Proc. Phys. Vol. 6 (Springer, Berlin, 1986).
25. R. Ulbricht, Elektr. Techn. Z. 26, 512 (1905).
26. R. Ulbricht, Elektr. Techn. Z. 28, 777 (1907).
27. J.A. Jacquez and H.F. Kuppelheim, J. Opt. Soc. Am. 45, 460 (1955).
28. J.G. Symons, J. Solar Energy Engr. 104, 251 (1982).
29. J.G. Symons, E.A. Christie and M.K. Peck, Appl. Opt. 21, 2827 (1982).
30. V.R. Weidner and J.J. Hsia, J. Opt. Soc. Am. 71, 856 (1981).
31. E. Krochmann and J. Krochmann, Paper presented at the CIE 20th session, 1983.
32. K.G.T. Hollands, G.D. Raithby and F.B. Russell, Int. J. Heat Mass Transfer 27, 2119 (1984).
33. K.G.T. Hollands and K. Iynkaran, Solar Energy 34, 309 (1985).
34. B. Jacobs, Diplomarbeit, University of Freiburg, Germany (1989).
35. B. Karlsson, in *Proc. North Sun 1988* (Swedish Council for Building Research, Sweden, 1988), p. 513.
36. J.G. Symons, J. Solar Energy Engr. 104, 251 (1982).
37. V. Wittwer, W. Stahl and A. Pflüger, Solar Energy Mater. 11, 199 (1984).
38. M. Rommel and V. Wittwer, in *Advances in Solar Energy Technology*, edited by W.H. Bloss and F. Pfisterer (Pergamon, Oxford, 1988), p. 641.
39. A. Goetzberger and M. Rommel, Solar Energy 39, 211 (1987).
40. C. Schmidt, A. Goetzberger and M. Rommel, in *Advances in Solar Energy Technology*, edited by W.H. Bloss and F. Pfisterer (Pergamon, Oxford, 1988), p. 935.

41. A. Goetzberger, J. Schmid and V. Wittwer, Arcus 1, 32 (1984).
42. A. Goetzberger, J. Schmid and V. Wittwer, Int. J. Solar Energy 2, 289 (1984).
43. W. Platzer, in *Advances in Solar Energy Technology*, edited by W.H. Bloss and F. Pfisterer (Pergamon, Oxford, 1988), p. 3498.
44. P.O. Braun, J. Schmid and E. Bollin, in *Proc. 7th Int. Sonnenforum*, Frankfurt (DGS-Sonnenenergie Verlags-GmbH, 1990), p. 544.
45. A. Goetzberger and K.A. Gertis, Technical Report BMFT-03E-8411-A (1987).
46. A. Goetzberger, Int. J. Solar Energy 2, 521 (1984).
47. A. Goetzberger and M. Rommel, Solar Energy 39, 211 (1987).
48. M. Rommel, V. Wittwer and A. Goetzberger, in *Advances in Solar Energy Technology*, edited by W.H. Bloss and F. Pfisterer (Pergamon, Oxford, 1988), p. 1553.

Chapter 4

SELECTIVELY SOLAR-ABSORBING SURFACE COATINGS: OPTICAL PROPERTIES AND DEGRADATION.

G.A. Niklasson and C.G. Granqvist

Physics Department
Chalmers University of Technology and University of Gothenburg
S-412 96 Gothenburg, Sweden

ABSTRACT

The most critical part of an energy-efficient solar collector is the absorber surface, which should be spectrally selective and exhibit high solar absorptance and low thermal emittance. This chapter introduces different design principles for achieving spectral selectivity, reports radiative properties for several available solar absorber surfaces, discusses theoretical modelling, and reviews recent work aimed at understanding their degradation at elevated temperature.

I. INTRODUCTION

An energy-efficient solar collector should absorb incident solar radiation, convert it to thermal energy, and deliver the thermal energy to a heat-transfer medium with minimum loss at each step. Figure 1 serves as a convenient introduction to the design of a flat-plate solar collector and outlines the most salient components. The collector comprises a thermally well-insulated arrangement whose upward-facing side is transparent so that solar radiation can penetrate to an absorbing surface, with carefully tailored properties, in contact with a heat-transfer medium such as water or air. Thermal losses are diminished by placing the absorber surface below a cover glass. Even smaller losses can be obtained by use of transparent insulation materials of the kind discussed in the previous chapter. In principle, the energy efficiency of the collector can be boosted by surface coated glass: antireflection coatings as well as infrared-reflecting coatings are of interest. Such coatings are discussed in the following chapter. It should be stressed that Fig. 1 refers to the commonly used fixed flat-plate collector. Most of the treatment below will be done with such collectors in mind, but we also include some discussion of absorber surfaces designed for high-temperature applications primarily in evacuated tubular collectors. Solar collector constructions with solar reflecting and tracking facilities are in existence as well.

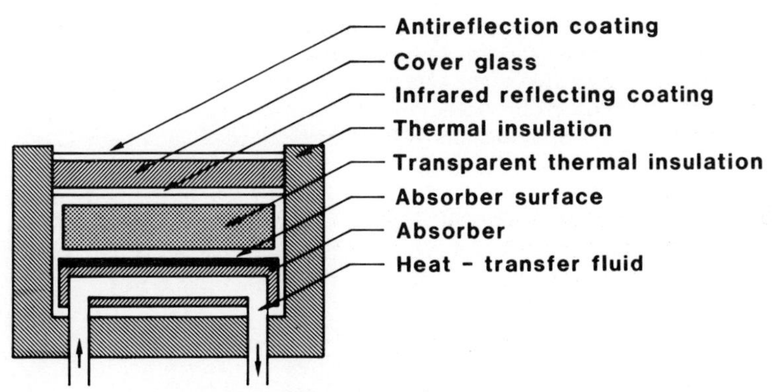

Fig. 1. Principle design of a flat-plate solar collector. Most practical collectors, though, do not use antireflection coatings, infrared reflecting coatings, or transparent thermal insulation materials. The different components are not to scale. (From Ref. 1).

The requirements for energy efficiency can be introduced with reference to the "natural" radiation in our surroundings. The pertinent radiative properties are shown in Fig. 2. The solid curve reproduces a typical spectrum for solar irradiance at the ground. Specifically, the curve gives the air mass (AM) 2 spectrum,[2] corresponding to clear weather with the sun 30° above the horizon. It is seen that solar radiation comes at wavelengths such that $\lambda < 3$ μm. The absorber surface of the solar collection device must absorb this energy. The surface then heats up and emits thermal radiation. The dashed curves in Fig. 2 indicate blackbody spectra for three temperatures; the emitted energy is negligible at $\lambda < 3$ μm for a temperature of $\tau < 100$°C. The losses associated with thermal emission should be avoided in order to gain energy efficiency. This can be accomplished in two different ways. The first one relies on a cover glass with a coating that reflects at $\lambda > 3$ μm so that the radiation emitted from the absorber surface is brought back to this same surface. A second, and more commonly used, way to diminish the heat losses is

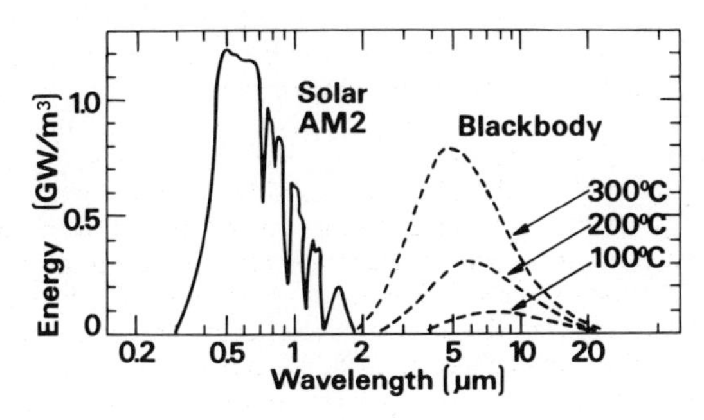

Fig. 2. Spectra for characteristic solar irradiance and for blackbody radiation pertaining to three temperatures.

to have an absorber surface whose thermal emittance is low. These are the selectively solar-absorbing surfaces to be discussed in this chapter. They are characterized by a low reflectance at $\lambda < \lambda_c$ and a high reflectance at $\lambda > \lambda_c$, where $\lambda_c \approx 3$ µm for $\tau < 100$°C. At increased operating temperatures, λ_c should be displaced towards a lower value, and at $\tau = 300$°C it is adequate to put $\lambda_c \approx 2$ µm.

Qualitative performance criteria can be formulated by use of the normal solar absorptance (A_{sol}) and the hemispherical thermal emittance (E_{therm}). These parameters are defined by

$$A_{sol} = \int d\lambda\, \phi_{sol}(\lambda)\,(1-R(\lambda, 0)) \Big/ \int d\lambda\, \phi_{sol}(\lambda), \quad (1)$$

$$E_{therm}(\tau) = \int d\lambda \int_0^{\pi/2} d(\sin^2\theta)\, \phi_{therm}(\lambda, \tau)\,(1-R(\lambda, \theta)) \Big/ \int d\lambda\, \phi_{therm}(\lambda, \tau), \quad (2)$$

$$\phi_{therm} = c_1\, \lambda^{-5} \left[\exp(c_2/(\lambda\tau))\right], \quad (3)$$

where ϕ_{sol} is the solar irradiance (for example the AM2 spectrum), $R(\lambda, \theta)$ is the reflectance as a function of wavelength and incidence angle, $c_1 = 3.7814 \times 10^{-16}$ Wm^{-2}, and $c_2 = 1.4388 \times 10^{-2}$ mK. Clearly, the desired spectral selectivity implies that A_{sol} should be close to unity and that E_{therm} should be minimized.

Coatings and surface treatments with high A_{sol} and low E_{therm} were subject to intense research and development in many laboratories around the world during the 1970's and early 1980's. This work has been reviewed in considerable detail in Refs. 3-12. An annotated bibliograpy,[13] covering the period 1955-1981, was published in 1983. It lists 565 scientific papers, including studies of almost 280 different coatings or surface treatments. Rather than reviewing this vast, and somewhat stale, field once again, this chapter focuses on certain key issues and on the latest developments. We first consider design principles for obtaining spectral selectivity, and data for several solar collector surfaces used in practice. Subsequently we discuss theoretical models for the optical properties and current ideas concerning degradation at elevated temperature. The treatment of degradation covers the latest advances for this technically important subject and is carried out in detail for some practically useful solar collector surfaces.

II. SELECTIVELY SOLAR-ABSORBING SURFACES: DESIGN AND DATA

A. Principles

It is possible to exploit several different design principles and physical mechanisms in order to create a selectively solar-absorbing surface. Six of these are shown schematically in Fig. 3. The most straightforward one is to use a material whose *intrinsic radiative properties* have the desired kind of spectral selectivity. Generally speaking, this approach has not been very fruitful, but work[14] on ZrB_2

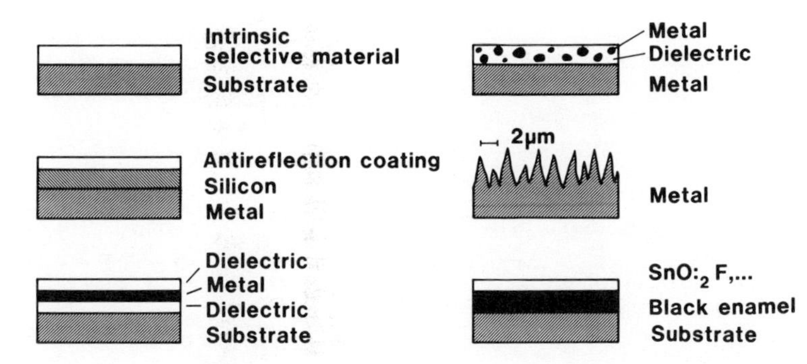

Fig. 3. Schematic designs of six different coatings and surface treatments for selective absorption of solar energy. (From Ref. 1).

and on some other compounds indicates that intrinsically selective materials do exist.

Semiconductor-metal tandems can give the desired spectral selectivity by absorbing short-wavelength radiation in a semiconductor whose bandgap is ~ 0.6 eV and having low thermal emittance as a result of the underlying metal. The useful semiconductors have undesirably large refractive indices, which tends to yield high reflection losses. Hence it is necessary to antireflect the surfaces in the range of solar radiation. A well known work[15,16] in this category is centered on Si-based designs prepared by chemical vapour deposition.

Multilayer absorbers can be tailored so that they become efficient selective absorbers of solar radiation. It is comparatively easy to compute their optical performance, which facilitates optimization of the design. One interesting example is $Al_2O_3/Mo/Al_2O_3$, which was originally developed for the U.S. space programme.[17] This type of surface has been produced by large-area sputtering technology.[18]

Metal-dielectric composite coatings consist of very fine metal particles in a dielectric host. The ensuing optical properties can be intermediate between those of the metal and of the dielectric. The effective medium theories introduced in the chapter on "Optical Properties of Inhomogeneous Two-Component Materials" are of great value for modelling the optical performance. The metal-dielectric concept offers a high degree of flexibility, and the optimization of the solar selectivity can be made with regard to the choice of constituents, coating thickness, particle concentration, and the size, shape, and orientation of the particles. The solar absorptance can be boosted by use of suitable substrate materials and antireflection treatments. The composite coatings can be produced by a variety of techniques such as electroplating, anodization, inorganic colouration of anodized aluminium, chemical vapour deposition and codeposition of metal and insulator by evaporation and sputtering. We return to several of these possibilities below.

Textured surfaces can produce a high solar absorptance by multiple reflections against metal dendrites that are ~2 μm apart, while the long-wavelength emittance is rather unaffected by this texture. Well-known examples are dendritic tungsten prepared by chemical vapour deposition[19] and textured copper, nickel, and stainless steel surfaces made by sputter etching.[20] Textured Al-Si (Ref. 21) will be discussed later.

The final concept considered here involves a *selectively solar-transmitting coating on a blackbody-like absorber*. The absorber can be chosen from among materials with proven long-term durability (such as black enamel), and the coating can be a heavily doped oxide semiconductor (for example SnO_2:F; cf. the following chapter). Coatings of this type are discussed in Ref. 22.

B. Results for Some Practically Useful Surfaces

The most widely used selectively solar-absorbing surface seems to be "black chromium", which is a complex composite of metallic chromium and dielectric Cr_2O_3. The metal concentration is low at the interface towards the air and increases with depth into the coating. The selective solar absorption appears to be a combination of the effect of the metal-dielectric composite and of the pronounced surface roughness that has been observed in these films.[23] Black chromium has a high void fraction and consists of rounded particles with sizes of about 100 nm.[24] The chromium crystallites are smaller, though, on the order of 10 nm, and are probably embedded in the oxide phase.[25] The inset in Fig. 4 gives a schematic picture of the microstructure. The original work on black chromium for solar energy applications was carried out by McDonald,[26] who modified the procedure for making decorative electroplated layers. Several subsequent studies have been made on the relationship between plating parameters and structural and optical properties; see, for example, Ref. 27. Spectral reflectance data from McDonald's initial study[26] are reproduced in the main part of Fig. 4. The figure also shows spectral reflectance[28] for two commercial electroplated black chromium coatings, specifically for products by Energie Solaire (ES) S.A. in Switzerland (coating backed by stainless steel) and by Mti Solar Inc. in the U.S. (coating backed by nickel-covered copper). For all of the coatings, the reflectance is low in the solar range and high in the thermal range, i.e., the spectral selectivity is large and approximately matches the ideal curve indicated by the dotted lines in Fig. 4. The ES coating has[28] $A_{sol} \approx 0.94$ and E_{therm} (100°C) ≈ 0.20, whereas the Mti coating has[28] $A_{sol} \approx 0.97$ and E_{therm} (100°C) ≈ 0.09. Figure 4 also shows some results[29] for smooth sputter-deposited Cr-Cr_2O_3 composite coatings.

Other commercially produced selectively solar-absorbing coatings comprise metallic Ni particles embedded in anodic Al_2O_3. The initial work was by Andersson et.al.[30] These coatings are made by a two-step process with an initial anodization of an aluminium sheet in, for example, dilute phosphoric acid. This transforms the surface layer of the metal into porous Al_2O_3. Subsequently metal is precipitated inside the pores by electrochemical means, for example by AC electrolysis in a bath containing nickel sulfate. Metal particle sizes are probably of the order of a few tens of nanometres. By modification of the second step one can precipitate particles of other metals. A detailed study of the relation between

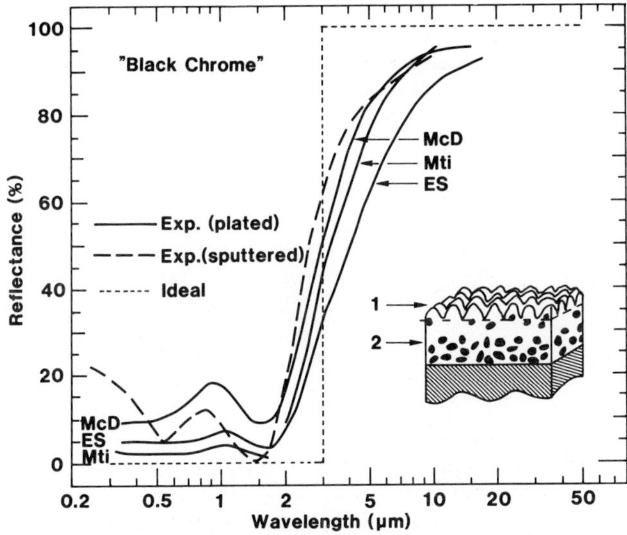

Fig. 4. Spectral reflectance measured for "black chromium" coatings on reflecting substrates. Solid curves refer to electroplated coatings prepared by McDonald (McD; Ref. 26), Mti Solar Inc. (Ref. 28), and Energie Solaire S.A. (ES; Ref. 28). The dotted curve refers to a sputter-deposited $Cr\text{-}Cr_2O_3$ composite coating (Ref. 29). Dotted lines represent an ideal reflectance profile. Inset indicates the microstructure for the electroplated coatings: a graded composite (layer 2) with a rough top surface (layer 1).

deposition conditions and spectral selectivity is reported in Ref. 31. In essence, the coating comprises a nickel-pigmented Al_2O_3 layer located under a porous Al_2O_3 layer. The porosity is largest at the outer surface. The bottom of the coating consists of a thin compact Al_2O_3 sheath serving as diffusion barrier. This microstructure is indicated in the inset of Fig. 5. The main part of the figure shows spectral reflectance for three types of $Ni\text{-}Al_2O_3$ coatings. Two of them are commercially produced by Sunstrip (SU) Viking AB in Sweden[28] and by Showa (SH) Aluminium Co. in Japan.[28,32] The third coating - denoted Gränges (GR) - was studied by Andersson et.al.[30] The SU coating has[28] $A_{sol} \approx 0.88$ and E_{therm} (100°C) ≈ 0.08, the SH coating has[28] $A_{sol} \approx 0.93$ and E_{therm} (100°C) ≈ 0.13, and the GR coating has[30] $A_{sol} \approx 0.92$ and E_{therm} (65°C) ≈ 0.10.

Recently Lanxner and Elgat reported[33] work on sputter-deposited molybdenum-based composite coatings for use on large-scale tubular solar collectors designed to operate at $\tau > 300$°C. The microstructure, indicated in the inset of Fig. 6, embodies two graded Mo-dielectric composite layers backed by an infrared-reflecting molybdenum layer. An antireflecting SiO_2 layer at the top and a diffusion barrier of Al_2O_3 at the bottom complete the design. Figure 6 shows that the reflectance has a high degree of spectral selectivity, with the steepest reflectance change around $\lambda \approx 2$ μm. It should be remarked that the main reflectance step should

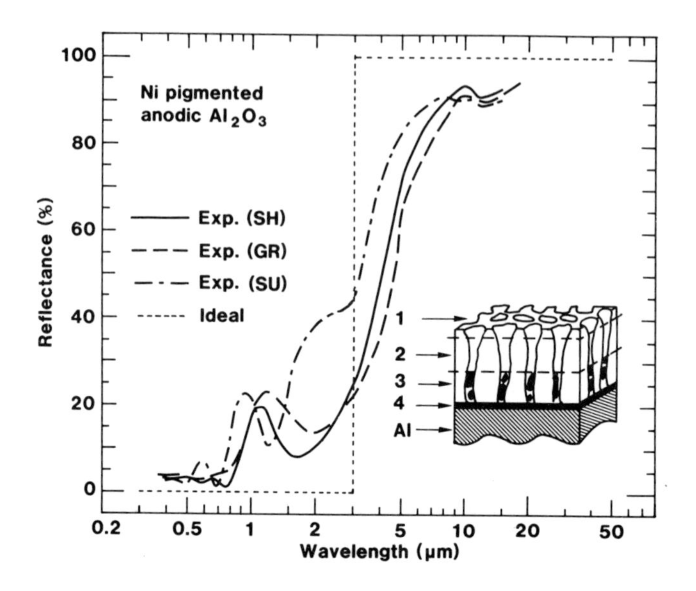

Fig. 5. Spectral reflectance measured for nickel-pigmented anodic Al_2O_3 coatings produced by electrochemical treatment of an aluminium sheet. Curves refer to coatings prepared by Gränges (GR; Ref. 30), Showa Aluminium Co. (SH; Refs. 28, 32) and Sunstrip Viking AB (SU; Ref. 28). Dotted lines represent an ideal reflectance profile. Inset indicates the microstructure of the coatings: An aluminium oxide diffusion barrier (layer 4), a Ni-Al_2O_3 composite (layer 3), and a porous aluminium oxide (layer 2) with increasing porosity towards the top surface (layer 1).

occur at a shorter wavelength for a coating devised to operate at a high temperature than for one to be used at ~100°C, which is easily inferred from Fig. 2. The coating in Fig. 6 is characterized by[33] $A_{sol} = 0.97 \pm 0.01$, E_{therm} (50°C) ≈ 0.10, and E_{therm} (350°C) $= 0.17 \pm 0.01$.

Among the remaining coatings that have reached commercialization, or for which processes adequate for practical manufacturing are known, we note chemically treated rough nickel surfaces (known under the trade name MAXORB[34]), "black nickel" made by electroplating,[35,36] graded stainless-steel carbon coatings developed for high performance tubular solar collectors,[37,38] copper oxide coatings,[39] and metal-filled coloured stainless-steel surfaces.[40] In this list we should also note Cr-Cr_2O_3 composite coatings made by "roll-coating" evaporation[41] and Al-Al_2O_3 made by "integral" colouration of anodic aluminium oxide.[42] Work on selectively solar-absorbing paints holds promise for extremely cheap surfaces.[43,44]

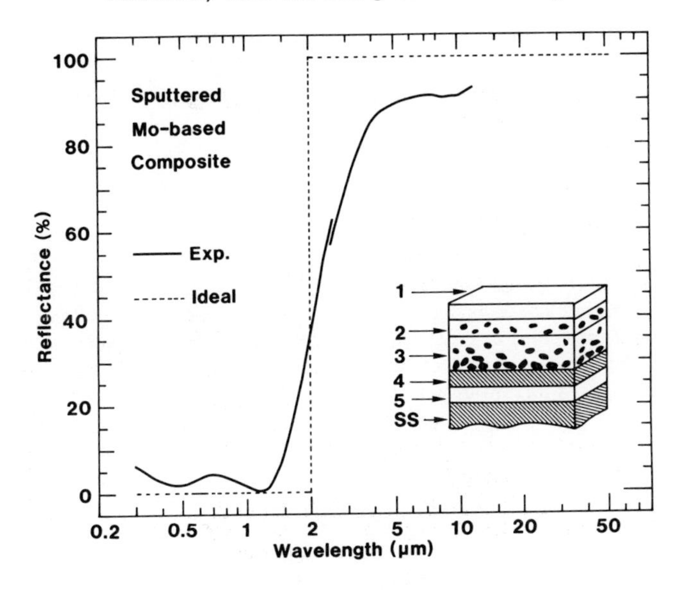

Fig. 6. Spectral reflectance measured for a molybdenum-based sputter-deposited coating (Ref. 33). Different recording methods were used for $\lambda < 2.5$ μm and $\lambda > 2.5$ μm, which explains the break in the curve. Dotted lines represent an ideal reflectance profile. Inset shows the coating microstructure with five superimposed layers. They comprise, from the top: SiO_2, graded Mo-SiO_2, graded Mo-Al_2O_3, molybdenum, and Al_2O_3. The substrate is of stainless steel (SS).

III. MODELS FOR MICROSTRUCTURE AND OPTICAL PROPERTIES

In the previous section we found that most commercially produced selectively solar-absorbing coatings consist of a mixture of metallic particles in a dielectric matrix and are backed by a metallic substrate. Commercial coatings are still largely produced by electrochemical methods (cf. Figs. 4 and 5), but physical vapour deposition techniques, such as sputtering, are becoming of increasing importance (cf. Fig. 6). In basic studies, however, physical vapour deposition has been widely used to extract detailed information on the relation between microstructural parameters of the coatings and their optical properties and spectral selectivity. In this section we consider microstructural models based on characterization studies of the pertinent coatings. With these models as a background, we discuss theories for the optical properties of selectively solar-absorbing surfaces with a metal-dielectric composite configuration.

A. Microstructure

It is suitable to make a distinction between metal-dielectric composities with homogeneous and graded composition. A homogeneous layer has the same structure throughout, while a graded layer displays a metal content that varies with the depth in the layer. In practice it is most advantageous to have a metal content that is high close to the substrate, and continuously decreases as one approaches the front surface.

A large amount of research has been carried out on metal-dielectric composite (also called "cermet") coatings, as can be seen in the bibliography.[13] A single composite layer on top of a metal substrate often does not give sufficient solar absorptance, and there are only a few cases where $A_{sol} > 0.90$ has been reported[45,46] for homogeneous composite layers backed by metal. Therefore more complicated designs have to be used. There are basically two ways to improve the optical properties of the coatings; they are shown schematically in Fig. 7. First, one may use antireflection coatings on top of the composite layer. This "Type I" design consists of, in order from the bottom, a metal substrate, a metal-dielectric composite layer, an antireflection layer, usually comprised of a dielectric, and a rough surface layer that may or may not be present. The outermost layer can be regarded as a composite of dielectric and air. This Type I configuration is often used in selective solar absorbers produced by evaporation and sputtering. The absorbing layer usually consists of a transition metal dispersed in an oxide, and it is convenient to produce the antireflection layer from the same oxide. Reported studies have concerned the materials Ni-SiO_2 (Ref. 47), Ni-MgO (Ref. 47), Cr-Cr_2O_3 (Ref. 29), Co-Al_2O_3 (Ref. 48), Pt-Al_2O_3 (Ref. 49) and Ni-Al_2O_3 (Ref. 50). In general, one can obtain $A_{sol} \approx 0.95$ and E_{therm} (100°C) ranging from 0.05 to 0.10. Among the commercially used coatings, it has been shown by Andersson et.al.[30] that the structure of metal pigmented anodic aluminium oxide solar absorbers is well described by the Type I configuration in Fig. 7. The model describes the optical properties of the coatings as well as the degradation of the solar absorptance at high temperatures as we will return to below.

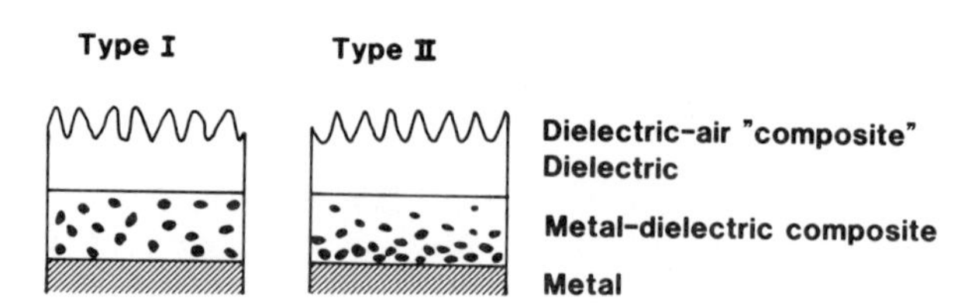

Fig. 7. Microstructural models for selectively solar-absorbing coatings.

In the second configuration (Type II in Fig. 7), the homogeneous metal-dielectric composite layer has been replaced by a graded one. Otherwise the design is similar to that of Type I. As before, the upper dielectric-air layer may or may not be present. In principle, the Type II structure can give a higher A_{sol} than the Type I structure, but in most cases the improvement is not very large. Model calculations and a discussion of optimum grading profiles are reported in Ref. 51. The use of graded metal-dielectric coatings for selective solar absorbers was reviewed by Craighead.[52] Coatings produced by evaporation and sputtering in most cases employ only a graded layer on a substrate, but in a few cases an antireflection layer is also included in the design.[53] Metal-dielectric composite layers have been produced from a wide range of transition and noble metals usually embedded in various oxides.[54,55] Among the more widely studied combinations, Pt-Al_2O_3 (Refs. 53, 56), Ni-Al_2O_3 (Ref. 56) and Ni-MgF_2 (Ref. 57) should be mentioned.

Chemically and electrochemically deposited coatings, such as the common commercial ones discussed above, display a very complex structure with graded or homogeneous composite layers, voids, oxide films, surface roughness, etc. This makes the modelling of their optical properties more difficult than for the sputtered and evaporated coatings, which are deposited in a more controlled way. However, the Type II configuration in Fig. 7 appears to be a suitable qualitative model for selective solar absorbers with a complex microstructure in many cases. In particular, Type II models have recently been used to model the optical properties as well as the degradation of the solar absorptance in a series of investigations by Scherer et.al. covering black zinc,[58] Ni-Al_2O_3 (Ref. 59) and copper oxide[60] coatings. Black chrome is realized to have a very complex structure (cf. Fig. 4) and grading models have been tried in order to describe its optical properties.[24,61,62] A very elaborate model by Sweet et.al.[24] could qualitatively describe features of the degradation behaviour for this kind of coating. This latter model employed a three-component mixture of chromium, Cr_2O_3 and air, where both the metal and void fractions were taken to be graded. The validity of this model is in question, though, as we will come back to below.

A further refinement of the microstructural models can be made by combining the Type I and Type II configurations so that the multilayer coating consists of, in order from the bottom, a metal substrate, a homogeneous composite layer, a graded composite layer, an antireflection coating, and, possibly, surface roughness modelled by an insulator-air grading. Unfortunately this kind of model has not been fully tested against experimental data, although the structure appears physically realistic. We will later present an example of a calculation of optical properties based on this model, which we call "Type III".

B. Applicability of Effective Medium Theory for the Optical Properties

The microstructural models in Fig. 7 form a suitable starting point for optical modelling by use of Effective Medium Theory (EMT). Several formulations of this type of theory were presented in the chapter on Optical Properties of Inhomogeneous Two-Component Materials above. The EMTs give the complex dielectric permeability of the composite material in terms of the permeabilities and volume fractions of the constituents, provided that the microstructure is specified in sufficient detail. In general, it is possible to reconcile experimental data and EMTs for selectively solar-absorbing coatings prepared by evaporation and sputtering. Calculations based on EMTs and the configurations in Fig. 7 lead to qualitative agreement between theory and experiment also for some chemically and electrochemically deposited coatings, but the fits are not always quantitative and may differ in detail. This is not surprising in view of the complexity of these latter coatings, for which the models in Fig. 7 should be regarded as approximations only.

It is appropriate to make some comments on the use of a graded composition profile to describe surface roughness effects by EMT. In these cases the profile of the dielectric permeability is derived from the composition profile by EMT, as proposed in Ref. 63. This procedure is questionable, though, unless the average period of the surface roughness is much less than the wavelength of light. The case of gratings with an arbitrary profile has been investigated rigorously.[64] Limits of validity for the grading model based on EMT have been derived for these gratings,[64,65] and it was shown that the wavelength-to-period ratio has to be larger than a number in the range of 5 to 40 in order that the quasistatic limit, that underlies the EMT, should be attained. The limit of validity of the EMT depends also on the height of the surface roughness.[65] This limit is evidently quite restrictive, and the surface roughnesses seen for example in electrochemically deposited selectively solar absorbing coatings is often of a larger magnitude. In these cases one should be cautious with the use of grading models. A conceptually better approach is provided by diffraction theory, which has been used to model the optical properties of black chromium coatings by Smith et.al.[23] Diffraction theory models the rough surface as a bigrating. It was shown that as-deposited black chromium exhibits a pronounced surface roughness with an average period well outside the quasistatic limit.[23] The diffraction theory yields optical properties in qualitative agreement with experiments and should be preferred over grading models for this coating. Unfortunately, restrictions on the height-to-period ratio that can be treated by current algorithms limit the applicability of diffraction theory for general rough surfaces.

C. Computational Procedures for Multilayered Coatings

We now briefly describe the methods used for calculations of the optical properties of coatings consisting of multilayered thin films, such as those of Fig. 7. Graded layers can also be treated as multilayers, i.e., the grading profile can be approximated by a large number of superimposed homogeneous layers. As a rule of thumb, we have found[66] that a good approximation is obtained by replacing the grading with 25 to 50 individual layers. The optical properties of a multilayer thin film can be calculated by standard methods (see, for example, Ref. 67). The basic steps are as follows:

(i) First the wavelength-dependent dielectric permeability $\varepsilon(\lambda)$ of the various layers needs to be known. For composite layers, EMT can be used for this purpose, provided that the dielectric permeabilities of the components and the composition are known. The most widely used EMTs are those of Maxwell Garnett[68] and Bruggeman.[69] For explicit expressions we refer to the earlier chapter. The permeabilities of the constituents, of the antireflecting coating, and the metal substrate may be obtained from independent measurements or from extensive tabulations that are readily available.[70,71]

(ii) This information, together with the thickness (d) of the individual layers, can now be used to obtain the optical response of the multilayer stack. The most convenient way is probably the method of characteristic matrices.[67] The characteristic matrix, **m**, of a single layer is written as

$$\mathbf{m} = \begin{pmatrix} m_{11} & m_{12} \\ m_{21} & m_{22} \end{pmatrix}, \tag{4}$$

where[67]

$$m_{11} = m_{22} = \cos\,[2\pi\, n(\lambda)\, d \cos\theta\, /\lambda] \tag{5}$$

and θ denotes the angle of incidence of the light. The complex refractive index $n(\lambda)$ is the square root of the dielectric permeability. The other components of the matrix are given by[67]

$$m_{12} = -i \sin\,[2\pi\, n(\lambda)\, d \cos\theta\, /\lambda]\, /P \tag{6}$$

and

$$m_{21} = -i\, P \sin\,[2\pi\, n(\lambda)\, d \cos\theta\, /\lambda]. \tag{7}$$

Here $P = \sqrt{\varepsilon/\mu}\cos\theta$ for s-polarised light and $P = \sqrt{\mu/\varepsilon}\cos\theta$ for p-polarised light, where μ is the magnetic permeability (which often can be set to unity). With knowledge of the characteristic matrices of the various layers in the stack, it is now fairly simple to obtain the resulting optical response. It can be shown[67] that the characteristic matrix of the multilayer stack, **M**, is simply the product of the matrices of the constituent layers.

(iii) We now consider a stack with layers numbered from 1 at the surface to N, denoting the substrate. Since we are presently considering metallic substrates, which are opaque, only the reflection from the system is non-zero. The reflection coefficient r of the multilayer film is readily obtained from[67]

$$r = \frac{(M_{11}+M_{12}P_N)P_1 - (M_{21}+M_{22}P_N)}{(M_{11}+M_{12}P_N)P_1 + (M_{21}+M_{22}P_N)}\ . \tag{8}$$

Here P_1 and P_N denote the values of the quanity P defined above for layers 1 and N, respectively. Finally, the reflectivity (R) is given by $R = |r|^2$.

IV. CASE STUDY ONE: OPTICAL PROPERTIES OF NICKEL PIGMENTED ALUMINIUM OXIDE

This section gives an example of the application of the structural multilayer models and the computational techniques for the optical properties discussed above. We present results from recent calculations[72] for various nickel pigmented aluminium oxide coatings on aluminium substrates. A later section will present models for their elevated temperature degradation. The manufacturing of the coatings was discussed briefly in connection with Fig. 5.

A. Model

A multilayer model for the optical properties, based on the most prominent structural features of the coating, was first advanced by Andersson et.al.[30] and has since been elaborated.[72] The structural model, depicted in the inset of Fig. 5, consists of four layers. They are numbered from the air interface to the substrate interface. Next to the aluminium substrate, a thin barrier layer of dense aluminium oxide is present. We will neglect this layer since it has only a minor influence on the optical properties.[30] Layer 3 consists of small nickel particles uniformly distributed in the porous oxide. Previously, it was shown[30] that good agreement with experiments can be obtained when the particles are taken to be spherical and the effective dielectric permeability of the composite is given by the Bruggeman EMT.[69] Layer 2 consists of porous oxide without metal particles. It is assumed that the porosity gradually increases towards the top of the stack, and in layer 1 we take a linear refractive index grading of aluminium oxide with air. This coating structure corresponds to Type I in Fig. 7.

The dielectric permeability of the composite layer was obtained from the permeabilities of the constituents by use of EMT. Specifically, the effective dielectric permeability, $\bar{\varepsilon} = \bar{\varepsilon}_1 + i\bar{\varepsilon}_2$, was given by the Bruggeman formula[69]

$$f_A(\varepsilon_A-\bar{\varepsilon})/(\varepsilon_A+2\bar{\varepsilon})+(1-f_A)(\varepsilon_B-\bar{\varepsilon})/(\varepsilon_B+2\bar{\varepsilon})=0, \tag{9}$$

where f_A is the volume fraction, or filling factor, of component A, and $\varepsilon_{A,B}$ denote the dielectric permeabilities of the constituents. One can use literature data for the wavelength-dependent dielectric permeabilities of nickel (Refs. 73-75), Al_2O_3 (Ref. 76), as well as for the aluminium substrate.[77,78]

The dielectric permeability of the porous oxide in layer 2 poses a problem. The porosity of anodized Al can be appreciable, typically up to 30 percent.[79,80] Therefore one must use a lower ε than that of homogeneous aluminium oxide.[76] For simplicity one may fix $\varepsilon = 2$ over the whole visible and near infrared wavelength range. The absorption of the oxide in this range is negligible. In the graded layer 4 it was assumed that ε decreases linearly from 2 to unity at the air interface.

B. Comparison of Theory and Experiment

Figure 8 compares calculated reflectance spectra to experiments for four nickel pigmented aluminium oxide coatings. Optical data for three of these were given in Fig. 5. The first test regards the Gränges (GR) coating, which appears to be the best characterized one. According to Andersson et.al.[30] one can use $d_3 = 0.3$ μm and $d_1 + d_2 = 0.4$ μm where d_i denotes the thickness of layer i. We consider a

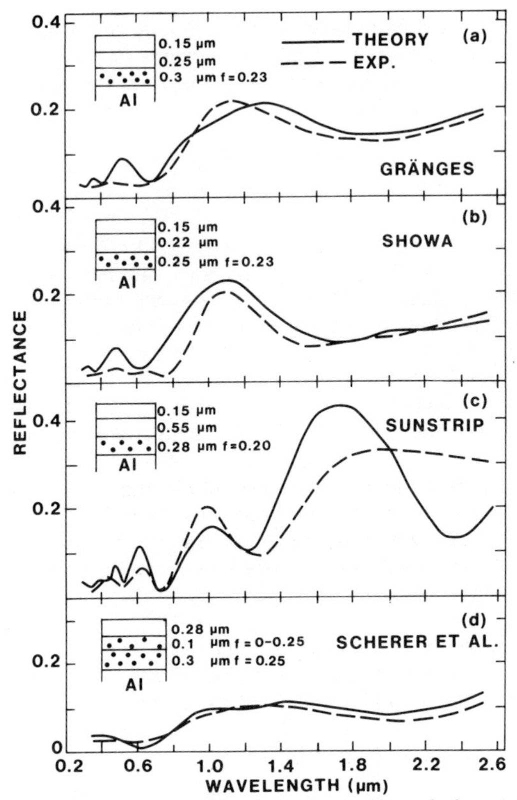

Fig. 8. Reflectance as a function of wavelength for nickel pigmented aluminium oxide coatings produced by (a) Gränges, (b) Showa, (c) Sunstrip and (d) Scherer et.al. Dashed curves show experimental data, and continuous curves show results of calculations using the structural models and parameter values shown in the insets. (From Ref. 72).

coating with a nickel density 0.62 g/m^2, which in combination with the value of d_3 yields a filling factor for nickel equal to f = 0.23. In the analysis we have varied the values of d_1 and d_2, keeping their sum constant, in order to obtain good agreement with the experimental spectrum. A calculation with the model presented above is compared to experimental data in Fig. 8a. We have used $d_1 = 0.15$ μm and $d_2 = 0.25$ μm. It is seen that the overall agreement is very good. Small discrepancies are present at $\lambda < 0.6$ μm, where the experimental reflectance is lower. This is probably due to a certain roughness of the interfaces between the various layers in the coating, as previously discussed.[30]

Figure 8b is based on the Showa (SH) coating. The close similarity between the optical properties of the Gränges and Showa coatings suggests that the parameters in the optical model should be rather similar. The final comparison with the experimental data,[28] which is presented in Fig. 8b, used the thickness values $d_1 = 0.15\ \mu m$, $d_2 = 0.22\ \mu m$ and $d_3 = 0.25\ \mu m$. The agreement is again very good except at the short wavelength end of the spectrum. The total thickness is consistent with electron microscopic studies.[81]

Before considering some other coatings we must make a short digression on the infrared properties of nickel pigmented aluminium oxide. Figure 9 presents a comparison of theory and experiment in the infrared wavelength range for the Gränges and Showa coatings. The model calculations account very well for the features in the spectra. At still longer wavelengths (above 10 μm), the optical properties are dominated by the lattice vibrations of aluminium oxide. These effects increase the thermal emittance of the coatings, especially of the heavily degraded ones. Modelling of optical spectra in this region is notoriously difficult since the phonon properties of the porous and probably impure anodic oxide cannot be expected to coincide with the phonons of thin films prepared under well-controlled conditions.

An important parameter for the nickel pigmented aluminium oxide coatings is the crossover wavelength λ^*, which is defined as the wavelength where the reflectance is 0.5. Empirically,[30] this wavelength depends linearly on the nickel

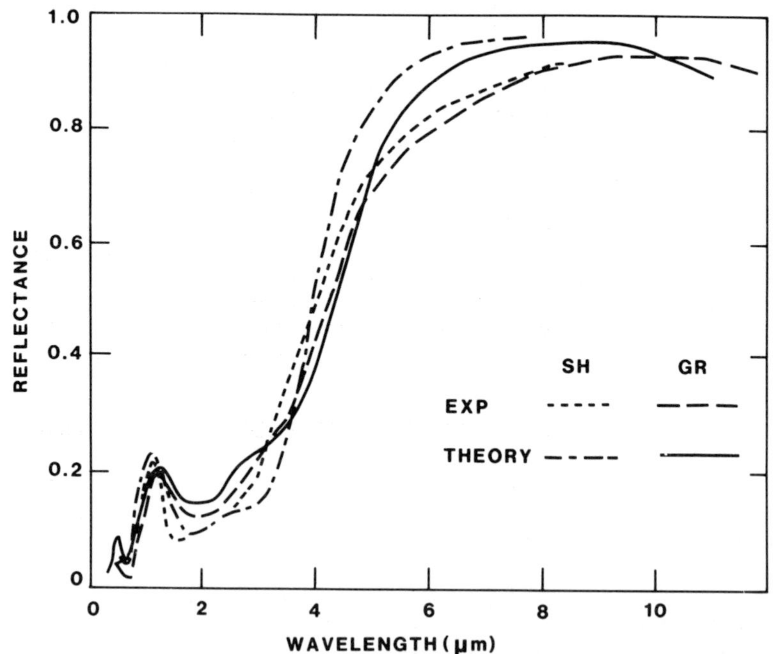

Fig. 9. Spectral reflectance for nickel pigmented aluminium oxide coatings produced by Showa (SH) and Gränges (GR). Experimental results are compared to calculations based on the models depicted in Figs. 8a and b. Each indicated curve is explained in the inset. (From Ref. 72).

content of the film. The computations showed that λ^* is primarily dependent on the nickel filling factor and, to a lesser degree, on the thickness of the composite layer.

We now consider the Sunstrip coating, for which very little structural information is presently available. For the data shown in Fig. 5, the crossover wavelength is $\lambda^* \approx 3$ to 3.5 μm, which suggests that the nickel filling factor is lower than for the Showa coating. According to Ref. 30, the shown value of λ^* corresponds to a nickel density of approximately 0.5 g/m^2. Several interference fringes are present in the experimental reflectance spectrum. A fit of the calculations to these fringes shows that the oxide thickness $d_1 + d_2$ should be in the range 0.6 to 0.7 μm. Figure 8c compares a computation with $f = 0.20$, $d_1 = 0.15$ μm, $d_2 = 0.55$ μm and $d_3 = 0.28$ μm to experimental data.[28] These parameter values are not unique, and equally good fits could be obtained with some other combinations. The agreement between theory and experiment was found to be reasonable at short wavelengths. Discrepancies appeared above 1.5 μm, where the calculation showed more pronounced interference oscillations than those present in the measured reflectance spectrum. One may speculate that this points to a minor inadequacy in the Bruggeman EMT when applied to these coatings.

Finally, some comments are given on the heavily pigmented coatings of Scherer et.al.[59] In this case one has to depart from the Type I models used above and instead apply the more complex Type III model to the structure of the coating. Indeed, it is necessary to insert a linearly graded nickel-aluminium oxide layer between layers 2 and 3 in order to obtain acceptable agreement with experiments. A comparison of calculations and experimental data[59] is given in Fig. 8d. The calculation employed a graded layer with $d_G = 0.1$ μm between $d_3 = 0.3$ μm and $d_2 = 0.28$ μm. The filling factor in layer 3 was put to 0.25, which is reasonable since this heavily coloured coating probably has a higher nickel content than the others. A graded oxide-air layer was not necessary in order to obtain the excellent agreement with the experiments shown in Fig. 8d. An almost equally good fit was obtained by using $d_1 = 0.15$ μm, as in the other cases, and at the same time decreasing d_2 to 0.2 μm. It is evident that heavily pigmented coatings can be described by a simple extension of the structural model shown in the inset of Fig. 5.

IV. CASE STUDY TWO: TEXTURED Al-Si COMPOSITE COATINGS

This section gives an example of a calculation based on a grading model for a selectively solar absorbing coating with a rough surface. We present results[21] for chemically etched metallic Al-Si composite films deposited on glass substrates.

Al-Si coatings, consisting of a mixture of small aluminium and silicon particles, were produced by simultaneous evaporation of the constituents from two electron-beam sources. The optical properties and dielectric permeability of as-deposited films were discussed elsewhere.[82,83] The films contained ~20 vol.% Si and had thicknesses of 1.2 μm. After deposition the films were etched in an ultrasonically stirred bath with 2.5 moles NaOH per dm^3 of deionized water.[21] As the etching proceeded, the appearance of the samples changed from shiny metallic to a yellow colour, and then became progressively darker. Etch times of approximately one minute were necessary in order to produce a black appearance.

Scanning electron microscopy showed that the etched films had very irregularly textured surface structures.[21] The lateral size of the protrusions seemed to be in the 20-50 nm range. From examination of the edge of cleaved samples, the texturing depth, h, appeared to be roughly 100-400 nm with the depth increasing with etch time.[21] We believe that the surface roughness is due to preferential etching of the aluminium phase. A schematic representation of the structure is shown in the inset of Fig. 10. The solid curve shows the experimental reflectance[21] for an Al-Si composite etched in NaOH for 55 s. The data correspond to $A_{sol} \approx 0.95$ and $E_{therm} \approx 0.2$.

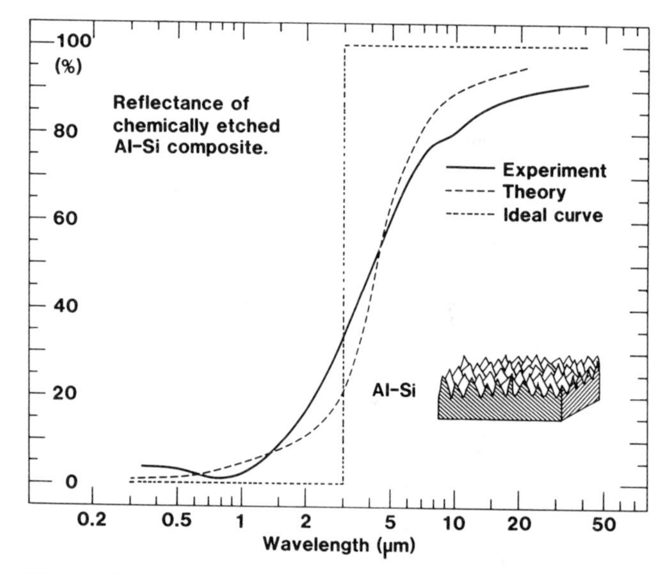

Fig. 10. Spectral reflectance for a chemically etched Al-Si surface. The dotted lines indicate the ideal reflectance, the solid curve represents experimental results obtained on an Al-Si composite etched in NaOH for 55 s, and the dashed curve shows the result of a computation based on a graded refractive index model employing textures 300 nm deep. (After Ref. 21).

The surface texture of the etched Al-Si films was modeled as a grading of the bulk Al-Si material with air. The texture was represented as half-spheroids subdivided into many layers. The refractive index profile was obtained from the Maxwell Garnett[68] EMT as applied to each layer. The filling factor profile was given by

$$f(z) = 1 - [(z - h)/h]^2, \qquad (10)$$

where the z-coordinate is zero at the outer surface of the graded layer and increases towards the substrate. We emphasize that grading models should be used with caution for this surface, since one cannot be sure that the quasistatic limit is attained at all wavelengths. However, the lateral size of the roughness is sufficiently small that the grading model based on EMT should be valid at least in the thermal infrared wavelength range. The dashed curve in Fig. 10 shows the result of a calculation with the model outlined above, using a grading depth of 300 nm. The agreement with the experimental curve is acceptable, despite the various simplifications made in the theoretical model.

We believe that this simple etch treatment can be applied to Al-Si surfaces produced by a more production-oriented method than electron-beam evaporation. Aluminium-silicon alloys are widely used in industry. Such alloys, of suitable microstructure, conceivably could be rendered spectrally selective by etching.

VI. DEGRADATION AND DURABILITY

The durability and degradation are of extreme importance for the technical application of selectively solar-absorbing coatings. The interest in degradation studies, and the determination of service life of solar absorbers from accelerated ageing tests, has increased rapidly during recent years.[84] Important studies on black chromium have been carried out by Pettit,[85] and several commercial coatings have been studied by Köhl et.al.[86,87] Studies of degradation and durability of solar collector coatings have been given a major impetus by the establishment of the International Energy Agency (IEA) Task X program.[28,88] This six-year collaborative effort, involving eleven countries, was begun in 1985.

A. Accelerated Ageing Tests

We first briefly review the most common accelerated ageing tests that have been applied to study the influence of various degradation factors on selectively solar-absorbing coatings.

(i) Ageing at elevated temperatures in air has been a common test.[28,84-87] It gives information on the performance and durability of a coating as a function of applied temperature.

(ii) The influence of atmospheric constituents on coatings can be studied by exposing them to increased concentrations of these substances or by increasing the temperature. Probably the most important of these degradation factors are humidity and SO_2. Accelerated humidity tests are often performed at 90°C and 95 % relative humidity,[87] but lower loads have also been used.[28]

(iii) During night the temperature of solar collectors may drop below the dew-point and consequently condensation can occur in the air gap between absorber and glazing. It is important that the durability of solar absorber coatings under these conditions be thoroughly tested.[87]

(iv) The influence of solar radiation can be tested by solar simulation lamps in the laboratory. The effect of UV-radiation appears to be especially important.[84]

(v) Of course, the various accelerated tests must be validated by comparison with results of outdoor exposures, which may be carried out under stagnation or operating conditions.

At present we focus on accelerated ageing tests in air at elevated temperatures. The reason for this is simply that the degradation processes in this case appear to be quite well understood and detailed models can be formulated. Two commercial selectively solar-absorbing coatings, will be considered below, namely nickel pigmented aluminium oxide and black chromium. This is done because these coatings were selected for detailed studies in the IEA Task X program, and a wealth of data is available.[28,88] It seems that a decrease of A_{sol} upon high temperature treatments is the main reason for the degradation of the performance of these solar absorber coatings. E_{therm} may either decrease or increase upon exposure to elevated temperatures, but the changes are ususally smaller and affect the performance less than the absorptance variations.[28] However, it should be noted that large increases of thermal emittance have been observed in nickel pigmented aluminium oxide absorbers upon condensation tests.[28,88]

B. Modelling of Optical Properties during Ageing

In order to develop some understanding of the underlying causes of the degradation, we model the changes of the optical properties during ageing[72] by use of the theoretical framework in the earlier sections. We first consider the aged nickel-aluminium oxide coatings and investigate the structural changes which lead to the observed changes in the optical spectra. After heat treatments in air at elevated temperatures, the optical properties of the Showa and Sunstrip coatings change in the same way. The main effect is a gradual increase in the reflectance at $\lambda > 1$ µm as the heat treatment progresses, as clearly seen in Fig. 11 which depicts data[28] for the Showa coating. Qualitatively similar results were obtained for Sunstrip.[28] This effect leads to a gradual lowering[28,87,88] of A_{sol}. Also a slight shift to shorter wavelengths of the peak at about 1 µm can be seen in Fig. 11. In the infrared one can notice that λ^* decreases,[28] which suggests that oxidation of the nickel particles is the main degradation mechanism.

The oxidation of the nickel in the coating can be described through two limiting models. One possibility is that the oxidation of nickel in layer 3 takes place uniformly throughout the layer. In this case the filling factor decreases as degradation proceeds, but d_3 remains constant. Model calculations for this case are shown in Fig. 12a with parameters pertinent to the Showa coating. The calculations reproduce the qualitative features of the experiments, except that the shift of the peak at ~1 µm appears to be larger in the experimental data. Further, the experiments show very flat reflectance curves at $\lambda > 1.5$ µm, while the calculated reflectance increases somewhat with wavelength. However, these discrepancies are only minor, and one can be confident that the degradation mechanism is a rather uniform oxidation of the nickel particles throughout the composite layer.

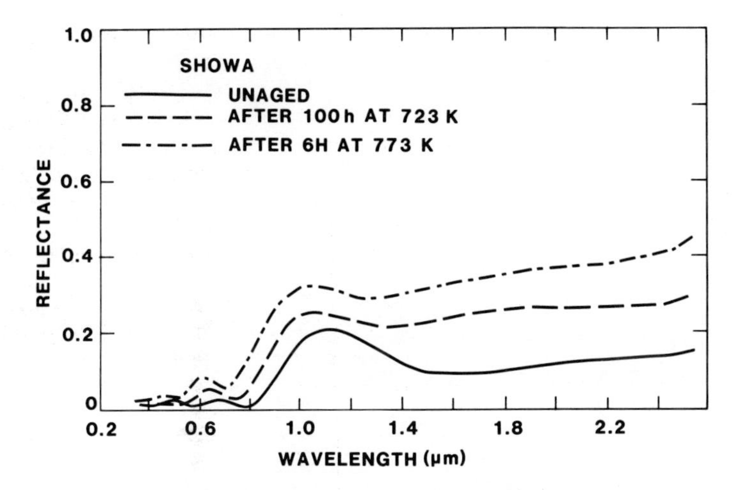

Fig. 11. Reflectance as a function of wavelength for nickel pigmented anodic aluminium oxide coatings produced by Showa. The continuous curve shows experimental results for an unaged coating, while dashed and dash-dotted curves show results for coatings aged for 100 h at 723 K and 6 h at 773 K, respectively. (From Ref. 28).

Secondly, an oxidation front may proceed into the coating, causing d_3 to gradually decrease and d_2 to increase as degradation continues. This possibility is explored in Fig. 12b, which presents calculations pertinent to the Showa coating with $f = 0.23$ and various values of d_3. The nickel content is given by the relative value Q, which is defined by

$$Q = W/W_0, \tag{11}$$

where W is the actual nickel content and W_0 denotes the initial nickel content. In Fig. 12b, Q is the ratio of the composite layer thickness after and before degradation. The values of Q were chosen so that they should be the same as in the calculations in Fig. 12a, where Q is given simply by the ratio of the actual filling factor to its initial value. It is seen in Fig. 12b that only slight changes in the reflectance spectra occur as Q decreases. A_{sol} actually increases slightly as Q is decreased. This is clearly in contradiction to the experimental results[28,87,88] and therefore the oxidation front model cannot explain the experimental spectra of heat treated samples. Calculations for the Sunstrip coating yielded the same qualitative features as those for Showa. Hence one must conclude that in this case also oxidation of nickel takes place uniformly in the composite layer.

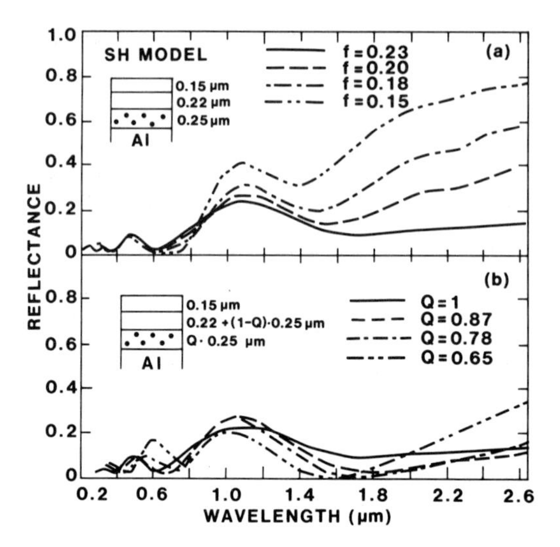

Fig. 12. Calculated reflectance as a function of wavelength for nickel pigmented aluminium oxide coatings. The calculations were carried out using parameters pertinent to the Showa coating. In (a) calculations for four different filling factors are shown. In (b) results for four different thicknesses of the composite layer and a filling factor of 0.23 are depicted. The parameter Q denotes the normalized nickel content in the film. The structural models and parameters used in the calculations, as well as the indicated curves, are shown in the insets. (From Ref. 72).

C. Parameterization of the Degradation

Having established the degradation mechanism as oxidation of nickel particles in the composite, we now turn to the question as to whether the degradation can be described by parameters in a simple way. To this end extensive calculations of A_{sol} were performed with parameters pertinent ot the Showa and Sunstrip samples. Equation (1) was used to obtain numerical data. For ϕ_{sol} we employed the air mass 1.5 spectrum, in accordance with the ASTM standard.[89] An interesting relationship was found between $1 - (A_{sol}/A^{0}_{sol})$ and the quantity Q defined above. Here A^{0}_{sol} denotes the solar absorptance of the coating prior to degradation.

Figure 13 shows a log-log plot of $1 - (A_{sol}/A^{0}_{sol})$ as a function of $1 - Q$. It is seen that the calculated points fall on a straight line both for the SH and SU models. Hence we infer the relationship

$$1 - (A_{sol}/A^{0}_{sol}) \sim (1 - Q)^{\beta}, \qquad (12)$$

which can be rewritten in terms of the change in solar absorptance, $\Delta A_{sol} = \overset{o}{A}_{sol} - A_{sol}$, as

$$\Delta A_{sol}/\overset{o}{A}_{sol} \sim ((f_0 - f)/f_0)^{\beta}, \tag{13}$$

for the case of degradation occurring by decrease of the filling factor in the composite layer. Hence we have found that the normalized solar absorptance change behaves as a power of the normalized change of nickel content. From Fig. 13 it is seen that $\beta \approx 2.5$ for the SH and SU models.

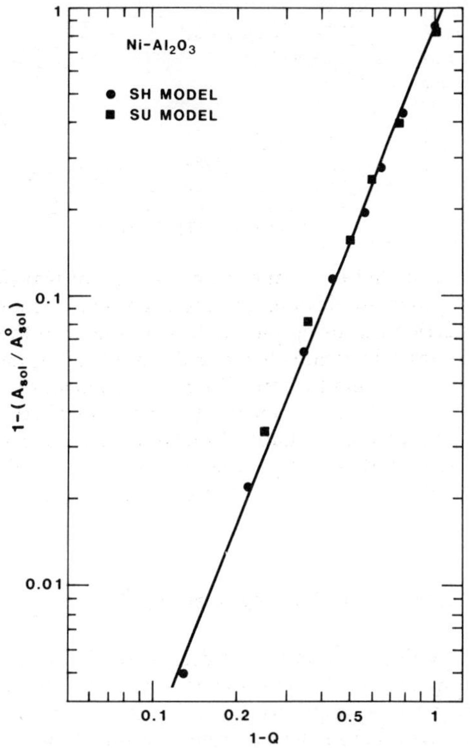

Fig. 13. Normalized solar absorptance change $(1 - (A_{sol}/\overset{o}{A}_{sol}))$ versus normalized decrease in nickel content (1-Q). Squares and circles denote calculations carried out with parameters pertinent to the Showa and Sunstrip coatings, respectively. The straight line represents Eq. (12) with $\beta = 2.5$. (From Ref. 72).

Model calculations for as-deposited and degraded samples similar to those above have not yet been carried out for black chromium coatings. As noted above, diffraction theory should be applicable to this problem, but the complexity of the calculations have so far prevented other than qualitative comparisons with experiments.[23] Also the degradation process appears to be complicated, involving both oxidation of chromium particles and a decrease in surface roughness.[23,90]

Figure 14 depicts literature data on A_{sol} of black chromium versus relative metal content, Q (Refs. 24, 90, 91). Only a few data points are available; they can be fitted to Eq. (12) with an exponent ß ~ 7. The curve extrapolates well to calculated A_{sol} (Ref. 28) for thin Cr_2O_3 films on chromium, and to experimental data.[92]

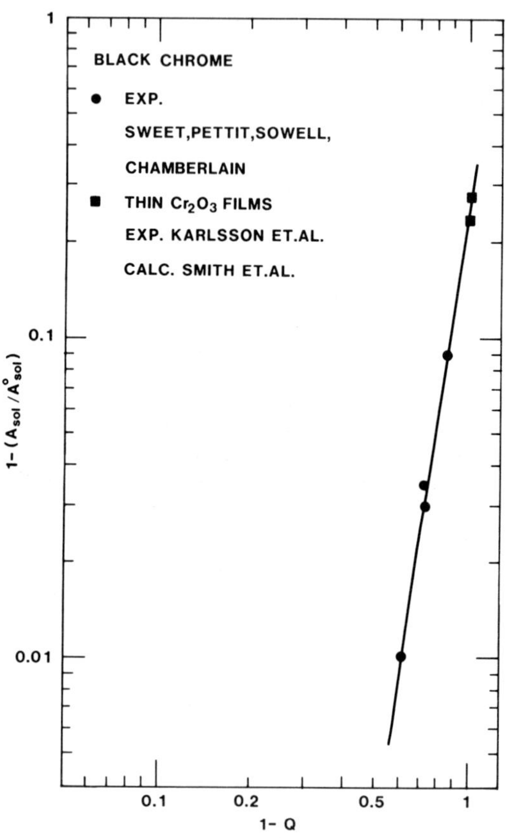

Fig. 14. Normalized solar absorptance change $(1 - (A_{sol}/A^{o}_{sol}))$ versus normalized decrease in chromium content (1 - Q) for black chromium coatings. Dots denote experiments of Pettit et.al. (Ref. 91) and Sweet et.al. (Ref. 24). Squares denote results for thin Cr_2O_3 films from Karlsson et.al. (Ref. 92) and Smith et.al. (Ref. 23). The straight line represents Eq. (12) with ß = 7.

So far we have only considered one of the parameters affecting the performance of a solar absorbing coating, namely A_{sol}. The magnitude of E_{therm}, considered in Ref. 28, seems to decrease slightly as degradation proceeds in most cases. This can be ascribed to the same mechanism as the decrease in A_{sol}. The increase of the reflectance at $\lambda > 1$ µm continues into the infrared range and causes E_{therm} to decrease gradually. On the other hand, as degradation progresses the amount of oxide in the coating increases. This leads to an increased infrared absorption due to the lattice vibration modes of the oxide. Hence this mechanism tends to increase E_{therm}. It is evident that modelling of the changes in emittance during degradation is a much more complex task than the modelling of the changes in solar absorptance considered above.

D. Oxidation Kinetics of Metals

The analysis above identified the oxidation of metal particles as the main reason for the elevated temperature degradation of some technically important selectively solar absorbing coatings. In order to progress towards a fundamental understanding of the degradation, it is important to have a clear notion of the current theories of the oxidation kinetics of metals. This subject is briefly reviewed below.

The oxidation kinetics of metals has been extensively studied for many years.[93-95] When a clean metal surface is exposed to oxygen, an initial fast oxidation occurs until a few monolayers have been formed. For the degradation of present interest, however, one is mainly concerned with the subsequent slow oxidation, which can be due to cation or anion diffusion assisted by an electric field that is set up across the oxide coating. The rate limiting step may be diffusion of ions, or transport of electrons in order to establish the electric field across the oxide. Oxides of the form MO, like NiO, can be characterized as so called network modifiers.[93] Here one expects cation transport to dominate the oxidation process. Xenon marker experiments[96] indicate that cation transport controls the oxidation of chromium as well. When the diffusion of the cations is the rate limiting step, the oxidation kinetics may be described by the theory of Cabrera and Mott.[97]

Cabrera and Mott assumed that adsorbed oxygen molecules on the surface dissociate and produce electron traps below the Fermi level of the metal. Electrons from the metal can then be transported through the oxide layer and fill these traps. Electron transport may be by tunnelling, or by thermionic emission for larger oxide thicknesses and higher temperatures. The electron transport leads to the formation of oxygen ions at the surface and hence an electric field is established in the oxide layer. It was assumed[93-95,97] that the potential drop, V, is independent of oxide thickness so that the field, F, is inversely proportional to the thickness. A very strong field can occur for thin oxide films. This electric field promotes the diffusion of cations through the oxide. When the diffusion of ions is the rate limiting step, the ion current, J_i, can be written in a simplified way as[93-95]

$$J_i = A\, N_o \nu \exp(-E_i/k\tau) \sinh(eaF/2k\tau), \qquad (14)$$

where A is approximately constant,[95] N_o is the number of metal ions per unit area on the metal/oxide interface, ν denotes a jump frequency, E_i is the activation energy for ion movement, e is the ion charge, a is the lattice constant and k is Boltzmann's constant. Equation (14) may be rearranged to yield the time dependent oxide thickness L(t) by[95]

$$dL(t)/dt = (L_c/t_c) \sinh (L_c/L(t)), \tag{15}$$

where $L_c = eaV/2k\tau$ and $t_c = (L_c/AN_o\nu v_i) \exp(E_i/k\tau)$, where v_i is the volume of oxide formed by a metal ion. Hence Eq. (15) describes the rate of metal oxidation by cation diffusion. Agreement of Eq. (15) with experiments does not establish unequivocally the physical mechanism, however. The rate equation for the case when thermionic emission of electrons is the rate limiting step leads to closely similar oxidation kinetics if the oxide film is sufficiently thin.[95] Indeed various assumptions regarding the physical mechanisms lead to a plethora of rate laws,[95,98] some of which can be closely approximated by Eq. (15).[98] For thick films ($L>>L_c$) the Cabrera-Mott law gradually crosses over to a diffusion limited behaviour characterized by a parabolic rate law.

The oxidation kinetics of nickel has been a subject of active study by many groups for the last fifty years.[99-105] The most detailed work appears to be the recent measurements of Sales et.al.[105] in the vicinity of the Curie temperature. Their experimental oxide thicknesses are depicted on a log-log plot as a function of time in Fig. 15. It is possible to let their data, obtained at different temperatures, fall on a single "master" curve by translation along the time axis. This is a consequence of the activated behaviour of t_c in Eq. (15). The location of the ten minute mark

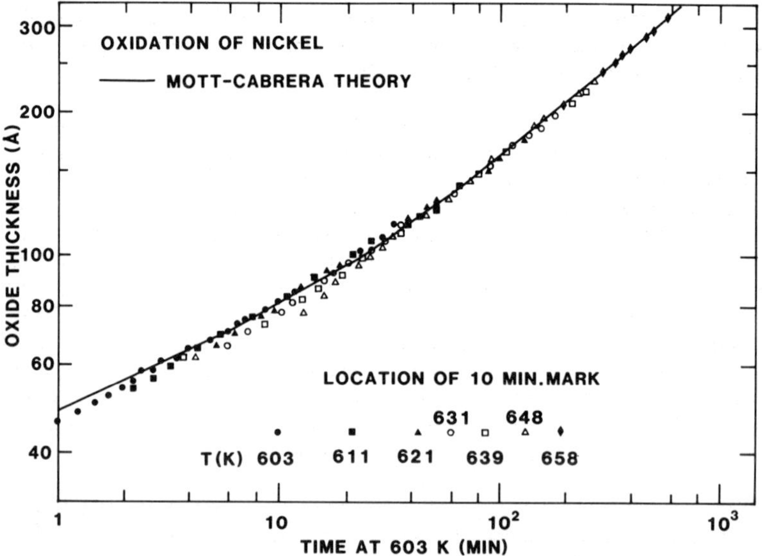

Fig. 15. Oxide thickness as a function of oxidation time for sintered nickel. Experimental points from Ref. 105, taken at different temperatures, have been superimposed on a single curve by translation along the time axis. The inset shows the location of the 10 minute mark for different temperatures. Symbols pertaining to all the temperatures are given in the inset. The curve depicts a prediction of the oxidation kinetics by the Cabrera-Mott theory. (From Ref. 72).

on the time axis is shown for each of the temperatures in Fig. 15. It is seen that the experimental points in the figure are in excellent agreement with the prediction of the Cabrera-Mott theory in Eq. (15). Hence one can conclude that the oxidation of nickel can be described with this theory, at least for oxide thicknesses below 30 nm. The qualitative features in Fig. 15 agree with earlier oxidation studies[101-103] for temperatures in the 650-800 K range. It should be noted that the oxidation kinetics in Fig. 15 can be approximated by the power law $L \sim t^{\phi}$ over regions of one decade or more in time. In the figure, ϕ is approximately 0.2 at short times and increases to 0.4 for longer times. At temperatures below 600 K, the oxidation kinetics of nickel can also be approximated by a power law. In this region, which corresponds to lower oxide thicknesses than those pertinent for Fig. 15, the exponent ϕ is lower. For example, values of ϕ between 0.06 and 0.13 have been found[99,100,104] at 470-475 K. These low exponent values are also in line with the Cabrera-Mott theory.

The activation energy of the oxidation process can be obtained from the shifts of the curves referring to different temperatures in Fig. 15. An equivalent procedure was used by Sales et.al.[105] to evaluate activation energies. They found a remarkable change in the activation energy at the Curie temperature of nickel (631 K). Below the Curie temperature the activation energy was found to be 2.65 eV, but the activation energy was only 1.60 eV at higher temperatures. Anomalous oxidation behaviour at the Curie temperature was seen also in other work,[107] where the oxidation rate showed a marked discontinuity at this temperature, but the activation energies differed only slightly.

Results on the oxidation of chromium at reduced oxygen pressure and temperatures in the range 573 to 673 K have also been fitted to the Cabrera-Mott theory, when the oxide thickness was below 6 nm (Ref. 106). The data can be well approximated by a power law with exponent 0.12. The activation energy was found to be 1.8 eV. At higher oxide thicknesses, the oxidation crosses over to a different regime with exponent 0.5. Hence it seems that the oxidation of chromium is more complex than for nickel, having different mechanisms competing. Shanker and Holloway[107] studied the oxidation of rough chromium surfaces in air for temperatures between 540 and 760 K. They reported a logarithmic rate law for oxide thicknesses up to 100 nm, but the results could also be fitted to power laws with kinetic exponents in the range 0.07 to 0.13. The oxidation rate was found to be dependent on sample preparation, and it was suggested that an increased surface roughness can decrease the oxidation rate.[107]

Very little is known about the oxidation behaviour of small metal particles. However, it has been shown that they can be passivated by slow oxidation[108,109] and are then remarkably resistant to further oxidation. Recently it was found[110] that the surface oxide on a metal particle is thinner the smaller the particle size. This effect was interpreted within the Cabrera-Mott theory as being due to the charging energy of the particles, which would prevent electron transport across the oxide and hence slow down the oxidation. Furthermore the roughness effect alluded to above indicates that small particles may display oxidation rates quite different from those of the bulk.

In order to apply the above considerations to the description of the oxidation of metal particles in solar collector coatings, one first needs some simple parameters that describe the oxidation kinetics. As noted above, a simple power law is a good approximation to the Cabrera-Mott theory for one to two decades of time. Hence we propose the relation

$$(1 - Q) \sim (t/q)^{\phi}, \quad (16)$$

where ϕ is the effective oxidation exponent referred to above, and q is assumed to be thermally activated according to the relation

$$q = q_0 \exp (E_a/k\tau), \quad (17)$$

where E_a denotes the activation energy. It should be noted that a similar relation has been previously used to analyze durability tests on solar absorber coatings by Pettit[85] and by Köhl et.al.[86.87]

E. Degradation Kinetics of Solar Absorber Coatings

In Sec. VI C it was found that the normalized solar absorptance change could be described as a power law of the normalized decrease in metal content. Analogously, from Sec. VI D one can infer that the oxidation kinetics of a metal often can be approximated by a power law over at least one to two decades of time. Based on these considerations, we propose a simple parametric equation that can be used to characterize and compare degradation curves of solar collector coatings. Combining Eqs. (12) and (16) one obtains

$$1 - (A_{sol}/A^{o}_{sol}) = B\,(t/q)^{\gamma}, \quad (18)$$

where B is a constant and the kinetic exponent is given by $\gamma = \beta\phi$. In recent years, many studies of the degradation kinetics of selectively solar absorbing coatings have been carried out. Here we analyze data obtained for nickel pigmented aluminium oxide and black chromium within the IEA Task X project.[28] In these studies, referred to above, A_{sol} was measured after annealing in air for various times at different temperatures.

In Fig. 16 we investigate to what extent Eq. (18) can describe the degradation of A_{sol} for selectively solar absorbing coatings upon exposure to elevated temperatures. In the plots the experimental data were normalized to a single curve by shifting the results obtained at various temperatures parallel to the time axis. Figure 16a depicts $1 - (A_{sol}/A^{o}_{sol})$ as a function of time, on a log-log plot, for the Sunstrip coating. The experimental points fall with a good approximation on a single line, which can be described by Eq. (18) with $\gamma = 0.17$. It is remarkable that the power law behaviour seems to persist for more than four decades on the time axis in this case. Figure 16b displays a similar plot for the degradation of a black chromium coating produced by Energie Solaire. Here γ shows a considerably higher value,

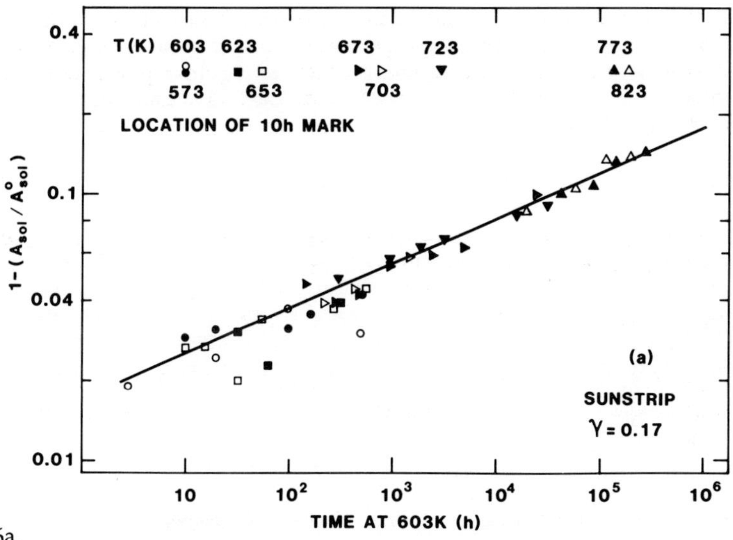

Fig. 16a.

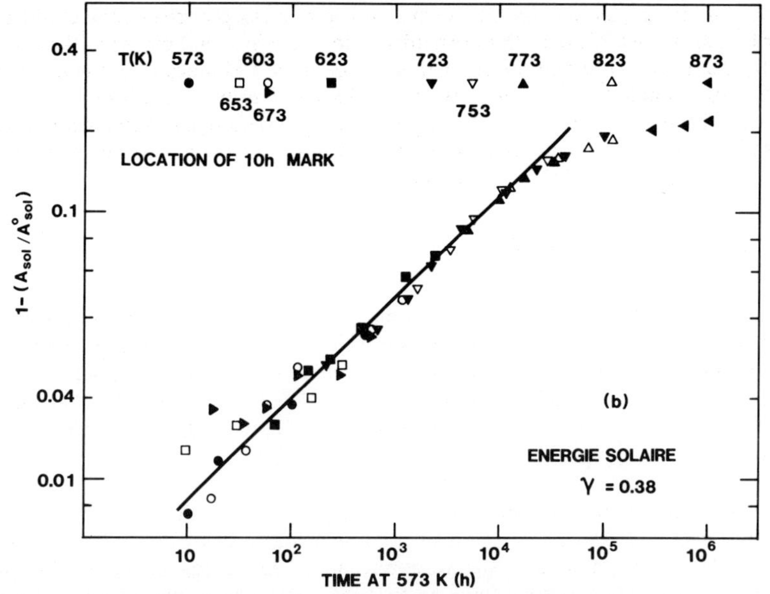

Fig. 16b. Normalized change in solar absorptance as a function of ageing time for (a) nickel pigmented aluminium oxide (Sunstrip) and (b) black chromium (Energie Solaire) coatings. Experimental data (Ref. 28) obtained at different temperatures were normalized to a single curve by shifting along the time axis, as indicated in the insets. Lines were drawn for convenience. (From Refs. 28 and 72).

which is in line with the larger values of ß (see Fig. 14) for black chromium. The activation energies for the degradation process can be determined from the shifts of the curves pertaining to different temperatures in Fig. 16 (see Eqs. 17 and 18). In Fig. 17 we depict the logarithm of the location of the 10 h mark, as given at the top of Fig. 16a, as a function of inverse temperature for the Sunstrip coating. The figure also contains data for the Showa coating. The activation energy was obtained from fits of the data points to straight lines as indicated in the figure. The scatter in the experimental points prevents an accurate determination of the activation energy, but data above the Curie temperature fall close to a straight line, and we obtain an activation energy of 2.4 eV. This value is much larger than the result of Sales et.al.[105] for the oxidation of bulk nickel. However, quite different activation energies have been found in different experiments. For example, values comparable with our result for the Sunstrip coating have been found[102] both below and above the Curie temperature in another study. A similar analysis, shown in Fig. 18, for the black chromium coating yields an activation energy of 1.5 eV, which is not far from the value for oxidation of bulk chromium.[106]

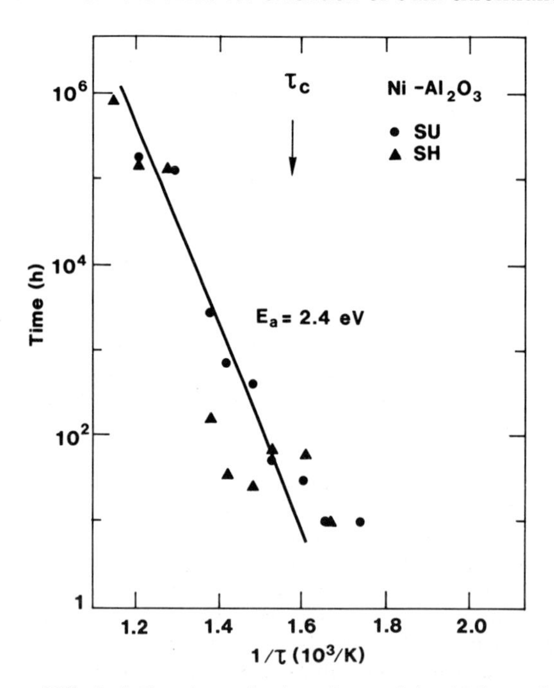

Fig. 17. Filled circles show the locations of the 10 h marks on the time axis of Fig. 16a as a function of inverse temperature for Sunstrip (SU) coatings. Filled triangles indicate analogous data for Showa (SH) coatings. A straight line fit to the data gave the shown activation energy. The Curie temperature of nickel is denoted τ_c. (From Ref. 72).

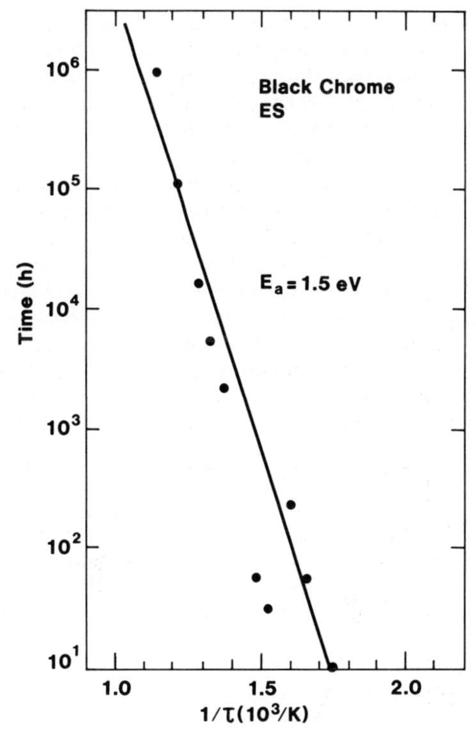

Fig. 18. Locations of the 10 h marks on the time axis as a function of inverse temperature for black chromium coatings made by Energie Solaire (ES). The results are based on an analysis similar to the one in Fig. 17. A straight line fit to the data gives the shown activation energy.

It is possible to make a preliminary comparison of the degradation kinetics of the solar absorber coatings with the oxidation behaviour of the bulk metals. From the exponent γ, describing the degradation kinetics, and the exponent ß, introduced in Sec. VI C, one easily obtains the kinetic exponent ϕ for the oxidation of metal in the selectively solar-absorbing coatings. The data presented above for nickel pigmented aluminium oxide and black chromium coatings lead to small values of ϕ, in the range of 0.05 to 0.07. This is lower than most of the results for bulk nickel and chromium. These low values of ϕ could be compatible with the Cabrera-Mott theory if the oxide coating is very thin. For the case of nickel oxide, thicknesses less than 3 nm are implied.

In addition, it follows from Fig. 13 that a change of 0.1 in $\Delta A_{sol}/A^o_{sol}$ corresponds to an oxide thickness of about one tenth of the metal particle diameter for the case of nickel pigmented aluminium oxide. Hence thicknesses of no more than a few nanometres are probably present in quite heavily degraded coatings.

There exists one major discrepancy between the oxidation of bulk metals and what can be inferred from Fig. 16. An oxide thickness of 5 nm is usually established in less than one hour at 600 K.[105,106] On the other hand, the degradation of $Ni\text{-}Al_2O_3$ and black chromium continues even after 10^4 hours at this temperature. Particles with sizes of a few tens of nanometres should oxidize completely in a much shorter time than this, if the bulk kinetics were obeyed.

The reason for the slow oxidation of the metal particles in selectively solar-absorbing coatings is not known and calls for fundamental studies. The oxidation mechanisms in small metal particles is largely an open question, and it is not clear whether the Cabrera-Mott theory is applicable at all, or if this theory can be modified to include particle size effects.

F. Lifetime Evaluations

Finally, we make some comments on lifetime evaluations for solar collector coatings based on accelerated ageing tests. The purpose of such tests is to obtain the same degradation of the sample as after a lifetime under operating conditions, but in a much shorter period. In high temperature degradation tests, this is accomplished by exposing the sample to higher temperatures than those usually encountered during normal operation of a solar collector. The so called acceleration factor[84-87] describes the difference in the degradation behaviour between the test temperature τ_{test} and a reference temperature τ_r (which represents, for example, normal operating conditions). It was shown in Sec. VI E that the degradation of the solar absorptance of nickel pigmented aluminium oxide and black chromium coatings is thermally activated. In this case the acceleration factor, a_{test}, is given by[84,87]

$$a_{test} = \exp\left[\frac{E_a}{k}\left(\frac{1}{\tau_r} - \frac{1}{\tau_{test}}\right)\right]. \tag{19}$$

In order to apply this equation to practical coatings, the following steps should be taken. (i) A criterion must be established, which states the decrease in solar absorptance that is considered acceptable for the performance of the solar absorber. (ii) Accelerated tests at different temperatures should be performed until this level of degradation is reached. (iii) The activation energy E_a should be calculated. (iv) Equation (19) should be used to calculate the acceleration factor and hence the lifetime of the selectively solar-absorbing coating under operating conditions.

Above, we confined the discussion to the case of degradation of A_{sol}, but acceleration factors can be used to describe the effect of various degradation modes[84,87] as well as combinations of them.[84] In these cases the functional forms of a_{test} may differ depending on the physics of the degradation modes.[84,87]

We also remark that Eq. (18), if validated for a particular selectively solar absorbing coating, permits the prediction of the future degradation of a sample from the degradation that has occurred in a certain time. However, it is not known whether functions of this form apply to degradation factors other than exposure to high temperatures.

VII. CONCLUSION AND REMARKS

This chapter introduced spectrally selective absorber surfaces for efficient photothermal conversion of solar energy. The key feature is a high solar absorptance combined with a low thermal emittance. The exposition is not complete but serves to update earlier reviews,[3-13] particularly as regards practically useful coatings, and to provide a detailed discussion on recent advances in understanding degradation and durability.

Several different design principles for achieving spectrally selective absorption were presented. In many cases, a metal-dielectric composite layer is responsible for the high solar absorptance, whereas the reflectance of an underlying metal substrate gives the low thermal emittance. The solar absorptance is often augmented by the use of antireflecting layers, porosity, surface roughness, etc. At least the majority of the practically useful selectively solar-absorbing surfaces of today have a metal-dielectric layer as their most crucial component. Many of the coatings - including black chromium and nickel pigmented anodic aluminium oxide - are prepared by electrochemical technology, but there appears to be a tendency to put more emphasis on coatings made by physical vapour deposition, such as magnetron sputtering. In our opinion, the sputter-deposited coatings made by LUZ Industries Israel - reported on in Fig. 6 above - represents a line of probable future development. The electrochemically produced coatings can be made with relatively inexpensive equipment, but require processing of large quantities of hazardous chemicals. Physical vapour deposition utilizes more capital-intensive equipment, but does not normaly lead to hazardous exhausts.

The different microstructures of relevance for selectively solar absorbing surfaces were introduced, and the pertinent Effective Medium Theories - needed for computing optical properties - were covered superficially. A more detailed exposition of these theories is given in the chapter on Optical Properties of Two-Component Materials earlier in this book. Multilayer configurations, surface roughness, and graded metal-dielectric compositions were discussed. Two case studies are included with the purpose of illustrating to what extent the optical properties can be understood from basic theory; specifically we discuss nickel pigmented anodic aluminium oxide and textured Al-Si composite coatings.

An important part of this chapter dealt with degradation and durability of some practically useful selectively solar-absorbing surfaces. The first part of the discussion on this subject regards modelling of the optical properties of different types of Ni-Al_2O_3 and black chromium surfaces subjected to accelerated degradation by heat treatment in air. From computations based on Effective Medium Theory it was shown that the degradation of Ni-Al_2O_3 progressed via oxidation of the nickel, so that the density of metallic nickel decreased uniformly across the metal pigmented layer. An alternative model, with an oxidation front that moves gradually toward the substrate during degradation, could not explain the data. In the next step of the analysis, the degradation was parameterized, and it was shown that the normalized solar absorptance behaves as a power law of the normalized nickel metal content. Well-defined values of the exponent in the power law could be extracted. The third step in the approach to a detailed understanding of the degradation involved a model for the oxidation kinetics of flat metallic surfaces. Here it is possible to apply the Cabrera-Mott theory,[97] which is known to be adequate for nickel and chromium, for example. Specifically, the oxidation kinetics can be represented by a power law with a fixed exponent over one to two decades in time. Small metallic particles have oxidation kinetics that are different from the kinetics of flat surfaces, though, and transform into oxide at a much slower rate. This phenomenon clearly is conducive to long-term durability. As a final step, the power law for the normalized change in solar absorptance was combined with the power law inherent in the Cabrera-Mott theory in order to formulate a parametric equation for the absorption degradation of selectively solar-absorbing coatings comprising a metal-dielectric composite. The parameters of this equation cannot yet be obtained from basic physics, which to a large degree is a manifestation of our lack of understanding of the oxidation kinetics of fine metal particles. It is hoped that continued work on solar collector durability, and on the durability of other metal-dielectric composites designed to operate at elevated temperature, will stimulate the needed experimental and theoretical work to understand the intriguingly slow oxidation of metal particles. The degradation kinetics of the thermal emittance is of interest for solar collector surfaces, but this effect has not yet been investigated in detail since it is much more difficult to treat than the absorptance degradation.

REFERENCES

1. C.G. Granqvist, *Spectrally Selective Surfaces for Heating and Cooling Applications* (SPIE Opt. Engr. Press, Bellingham, 1989).
2. P. Moon, J. Franklin Inst. 230, 583 (1940).
3. A.B. Meinel and M.P. Meinel, *Applied Solar Energy: An Introduction* (Addison-Wesley, Reading, 1976), Chap. 9.
4. B.O. Seraphin and A.B. Meinel, in *Optical Properties of Solids - New Developments*, edited by B.O. Seraphin (North-Holland, Amsterdam, 1976), Chap. 17.
5. R.E. Hahn and B.O. Seraphin, Phys. Thin Films 10, 1 (1978).
6. B.O. Seraphin, in *Solar Energy Conversion: Solid State Physics Aspects*, Vol. 31 of *Topics in Applied Physics*, edited by B.O. Seraphin (Springer, Berlin, 1979), p. 5; in *Solar Energy Conversion: An Introductory Course*, edited by A.E. Dixon and J.D. Leslie (Pergamon, New York, 1979), p. 287.
7. C.M. Lampert, Solar Energy Mater. 1, 319 (1979); 2, 1 (1979).
8. P.K.C. Pillai and R.C. Agarwal, Phys. Stat. Sol. A 60, 11 (1980).
9. O.P. Agnihotri and B.K. Gupta, *Solar Selective Surfaces* (Wiley, New York, 1981).
10. M.M. Koltun, *Selektivnye Opticheskie Poverknosti Preobrazovatelei Solnechnoi Energii* (Nauka Press, Moscow, 1979). English translation: *Selective Optical Surfaces for Solar Energy Converters* (Allerton Press, New York, 1981).
11. S.A. Herzenberg and R. Silberglitt, Proc. Soc. Photo-Opt. Instrum. Engr. 324, 92 (1982).

12. W.F. Bogaerts and C.M. Lampert, J. Mater. Sci. 18, 2847 (1983).
13. G.A. Niklasson and C.G. Granqvist, J. Mater. Sci. 18, 3475 (1983).
14. E. Randich and D.D. Allred, Thin Solid Films 83, 393 (1981); E. Randich and R.B. Pettit, Solar Energy Mater. 5, 425 (1981).
15. B.O. Seraphin, Thin Solid Films 39, 87 (1976); 57, 293 (1979).
16. M. Janai, D.D. Allred, D.C. Booth and B.O. Seraphin, Solar Energy Mater. 1, 11 (1979); D.C. Booth, D.D. Allred and B.O. Seraphin, Solar Energy Mater. 2, 107 (1979).
17. R.N. Schmidt and K.C. Park, Appl. Opt. 4, 917 (1965).
18. J.A. Thornton and J.L. Lamb, Thin Solid Films 96, 175 (1982).
19. G.D. Pettit, J.J. Cuomo, T.H. Di Stefano and J.M. Woodall, IBM J. Res. Dev. 22, 372 (1978).
20. G.L. Harding and M.R. Lake, Solar Energy Mater. 5, 445 (1981).
21. G.A. Niklasson and H.G. Craighead, J. Appl. Phys. 54, 5488 (1983).
22. F. Simonis, M. van der Leij and C.J. Hoogendoorn, Solar Energy Mater. 1, 221 (1979).
23. G.B. Smith, R.C. McPhedran and G.H. Derrick, Appl. Phys. A 36, 193 (1985).
24. J.N. Sweet, R.B. Pettit and M.B. Chamberlain, Solar Energy Mater. 10, 251 (1984).
25. C.M. Lampert and J. Washburn, Solar Energy Mater. 1, 81 (1979).
26. G.E. McDonald, Solar Energy 17, 119 (1975).
27. K.J. Cathro, Metal Finishing 76, (10), 57 (1978).
28. B. Carlsson, editor, *Accelerated Life Testing of Solar Energy Materials: Case Study of some Selective Solar Absorber Coatings for DHW-systems*, IEA SHC Technical Report (to be published).
29. J.C.C. Fan and S.A. Spura, Appl. Phys. Lett. 30, 511 (1977).
30. Å. Andersson, O. Hunderi and C.G. Granqvist, J. Appl. Phys. 51, 754 (1980).
31. M. Uchino, S. Aso, S. Hozumi, H. Tokumasu and Y. Yoshioka, Matsushita Electr. Ind. Natl. Techn. Rep. 25, 994 (1979).
32. S. Tsuda and Y. Asano, Belg.-Ned. Tijdschr. Oppervlaktetechn. Met. 22, 3 (1978).
33. M. Lanxner and Z. Elgat, Proc. Soc. Photo-Opt. Instrum. Engr. 1272, 240 (1990).
34. J.J. Mason and T.A. Brendel, Proc. Soc. Photo-Opt. Instrum. Engr. 324, 139 (1982).
35. H.Y.B. Mar, R.E. Peterson and P.B. Zimmer, Thin Solid Films 39, 95 (1976).
36. P.K. Gogna, K.L. Chopra and S.C. Mullick, Energy Res. 4, 317 (1980).
37. S. Craig and G.L. Harding, Thin Solid Films 101, 97 (1983).
38. Z.-C. Yin, private communication.
39. A. Roos, T. Chibuye and B. Karlsson, Solar Energy Mater. 7, 453 (1983); A. Roos and B. Karlsson, Solar Energy Mater. 7, 467 (1983).
40. J.J. Mason, Proc. Soc. Photo-Opt. Instrum. Engr. 428, 159 (1983).
41. ULVAC Corp., Tokyo, Japan.
42. W.C. Cochran and J.M. Powers, Aluminium (Düsseldorf) 54, 147 (1978).
43. S.W. Moore, Proc. Soc. Photo-Opt. Instrum. Engr. 324, 148 (1982).
44. B. Orel, I. Radoczy and Z. Crnjak Orel, Solar and Wind Technol. 3, 45 (1986); Z. Crnjak Orel, B. Orel and A. Krainer, Proc. Soc. Photo-Opt. Instrum. Engr. 1272, 274 (1990).

45. U.Kh. Gaziev, Sh.A. Faiziev, V.V. Li and V.S. Trukhov, Geliotekh. 16, 30 (1980) [Appl. Solar Energy 16, 30 (1980)].
46. F. Garnich and E. Sailer, Solar Energy Mater. 20, 81 (1990).
47. M. Okuyama, K. Furusawa and Y. Hamakawa, Solar Energy 22, 479 (1979).
48. G.A. Niklasson and C.G. Granqvist, J. Appl. Phys. 55, 3382 (1984).
49. T.K. Vien, C. Sella, J. Lafait and S. Berthier, Thin Solid Films 126, 17 (1985).
50. T.S. Sathiaraj, R. Thangaraj and O.P. Agnihotri, Solar Energy Mater. 18, 343 (1989).
51. I.T. Ritchie and B. Window, Appl. Opt. 16, 1438 (1977); B. Window, Solar Energy Mater. 2, 395 (1980).
52. H.G. Craighead, Proc. Soc. Photo-Opt. Instrum. Engr. 401, 356 (1983).
53. J.A. Thorton and J.L. Lamb, Thin Solid Films 83, 377 (1981); Solar Energy Mater. 9, 415 (1984).
54. R.C. Bastien, R.R. Austin and T.P. Pottenger, Proc. Soc. Photo-Opt. Instrum. Engr. 140, 140 (1978).
55. D.R. McKenzie, Appl. Phys. Lett. 34, 25 (1979); Thin Solid Films 62, 317 (1979).
56. H.G. Craighead and R.A. Buhrman, J. Vac. Sci. Technol. 15, 269 (1978).
57. M. Mast, K. Gindele and M. Köhl, Thin Solid Films 126, 37 (1985).
58. A. Scherer and O.T. Inal, Appl. Opt. 24, 3348 (1985).
59. A. Scherer, O.T. Inal and R.B. Pettit, J. Mater. Sci. 23, 1934 (1988).
60. A. Scherer, O.T. Inal and R.B. Pettit, J. Mater. Sci. 23, 1923 (1988).
61. I.T. Ritchie, S.K. Sharma, J. Valignat and J. Spitz, Solar Energy Mater. 2, 167 (1979/1980).
62. G. Zajac, G.B. Smith and A. Ignatiev, J. Appl. Phys. 51, 5544 (1980).
63. R.B. Stephens and G.D. Cody, Thin Solid Films 45, 19 (1977).
64. J.M. Bell, G.H. Derrick and R.C. McPhedran, Opt. Acta 29, 1475 (1982).
65. R.C. McPhedran, L.C. Botten, M.S. Craig, M. Nevière and D. Maystre, Opt. Acta 29, 289 (1982).
66. G.A. Niklasson, unpublished results.
67. M. Born and E. Wolf, *Principles of Optics,* 6th edition (Pergamon, Oxford, 1980).
68. J.C.M. Garnett, Philos. Trans. R. Soc. (London) 203, 385 (1904); 205, 237 (1906).
69. D.A.G. Bruggeman, Ann. Phys. (Leipzig) 24, 636 (1935).
70. J.M. Weaver, C. Krafka, D.W. Lynch, and E.E. Koch, *Physics Data: Optical Properties of Metals* (Fachinformationszentrum Energie, Physik, Mathematik GmbH, Karlsruhe, 1981), Vols. 18-1 and 18-2.
71. E.D. Palik, editor, *Handbook of Optical Constants of Solids* (Academic, New York, 1985).
72. G.A. Niklasson, Proc. Soc. Photo-Opt. Instrum. Engr. 1272, 250 (1990).
73. P.B. Johnson and R.W. Christy, Phys. Rev. B 9, 5056 (1974).
74. A.P. Lenham and D.M. Treherne, in *Optical Properties and Electronic Structure of Metals and Alloys,* edited by F. Abeles, (North-Holland, Amsterdam, 1966), p. 196.
75. A.S. Siddiqui and D.M. Treherne, Infrared Phys. 17, 33 (1977).
76. T.S. Eriksson, A. Hjortsberg, G.A. Niklasson and C.G. Granqvist, Appl. Opt. 20, 2742 (1981).
77. A.G. Mathewson and H.P. Myers, Phys. Scripta 4, 291 (1971).
78. H.J. Hagemann, W. Gudat and C. Kunz, DESY SR-74/7, May 1974.
79. T. Pavlovic and A. Ignatiev, Thin Solid Films 138, 97 (1986).

80. A. Scherer and O.T. Inal, Thin Solid Films 101, 311 (1983).
81. P.R. Dolley and M.G. Hutchins, in *Solar Optical Materials: Proc. Conf. Oxford, U.K., 12-13 April 1988,* edited by M.G. Hutchins, (Pergamon, Oxford, 1988), p. 91.
82. G.A. Niklasson and H.G. Craighead, Appl. Opt. 22, 1237 (1983).
83. G.A. Niklasson, D.E. Aspnes and H.G. Craighead, Phys. Rev. B 33, 5363 (1986).
84. B. Carlsson, editor, *Solar Materials Research and Development: Survey of Service Life Prediction Methods for Materials in Solar Heating and Cooling* (Swedish Council for Building Research, Stockholm, 1989).
85. R.B. Pettit, Solar Energy Mater. 8, 349 (1983).
86. M. Köhl, K. Gindele and M. Mast, Solar Energy Mater. 16, 155 (1987).
87. M. Köhl, K. Gindele, U. Frei and T. Häuselmann, Solar Energy Mater. 19, 257 (1989).
88. M.G. Hutchins, P.R. Dolley, K. Gindele, M. Köhl, U. Frei, B.O. Carlsson, S. Tanemura, K.G.T. Hollands, A.J. Faber, P.A. van Nijnatten and E. Mezquida, Proc. Soc. Photo-Opt. Instrum. Engr. 1016, 279 (1988).
89. ASTM Standard E 891-82, *Annual Book of ASTM Standards,* Vol. 12.02 (1986).
90. P.H. Holloway, K. Shanker, R.B. Pettit and R.R. Sowell, Thin Solid Films 72, 121 (1980).
91. R.B. Pettit, J.N. Sweet and R.R. Sowell, in *American Society for Testing and Materials*, Spec. Techn. Publ. No. 792, p. 263 (1983).
92. B. Karlsson, T. Karlsson and C.-G. Ribbing, Proc. Soc. Photo-Opt. Instrum. Engr. 324, 156 (1982).
93. F.P. Fehlner and N.F. Mott, Oxidation of Metals 2, 59 (1970).
94. K.R. Lawless, Rep. Progr. Phys. 37, 231 (1974).
95. A.T. Fromhold, Jr., *Theory of Metal Oxidation,* (North-Holland, Amsterdam, 1976), Vols. 1 and 2.
96. G. Salomonsen, N. Norman, O. Lønsjø and T.G. Finstad, J. Phys. Condens. Matter, 1, 7843 (1989).
97. N. Cabrera and N.F. Mott, Rep. Progr. Phys. 12, 163 (1948-49).
98. D.J. Young and M.J. Dignam, J. Phys. Chem. Solids 34, 1235 (1973).
99. W.E. Campbell and U.B. Thomas, Trans. Electrochem. Soc. 91, 623 (1947).
100. W. Scheuble, Z. Phys. 135, 125 (1953).
101. H.-J. Engell, K. Hauffe and B. Ilschner, Z. Elektrochem. 58, 478 (1954).
102. H. Uhlig, J. Pickett and J. Macnairn, Acta Met. 7, 111 (1959).
103. K. Hauffe, L. Pethe, R. Schmidt and S. Roy Morrison, J. Electrochem. Soc. 115, 456 (1968).
104. M.J. Graham and M. Cohen, J. Electrochem. Soc. 119, 879 (1972).
105. B.C. Sales and M.B. Maple, Phys. Rev. Lett. 39, 1636 (1977); B.C. Sales, M.B. Maple and F.L. Vernon III, Phys. Rev. B 18, 486 (1978).
106. D.J. Young and M. Cohen, J. Electrochem. Soc. 124, 769 (1977).
107. K. Shanker and P.H. Holloway, Thin Solid Films 105, 293 (1983).
108. K. Haneda and A.H. Morrish, Nature 282, 186 (1979).
109. A. Johgo, E. Ozawa, H. Ishida and K. Shoda, J. Mater. Sci. Lett. 6, 429 (1987).
110. S. Sako, K. Ohshima and T. Fujita, J. Phys. Soc. Japan 59, 662 (1990).

Chapter 5

ENERGY-EFFICIENT WINDOWS: PRESENT AND FORTHCOMING TECHNOLOGY

C.G. Granqvist

Physics Department
Chalmers University of Technology and University of Gothenburg
S-412 96 Gothenburg, Sweden

ABSTRACT

Current research and development offers important opportunities for improved energy efficiency of architectural windows. This chapter covers the design criteria for different climates and reviews means to fulfill these criteria by proper materials selection. We discuss glass properties, the importance of multiple glazing, the many uses of surface coated glass, and possibilities connected with materials interposed between glass panes. Among the coatings, we treat those based on noble metals and on doped oxide semiconductors, both of which have static spectrally selective properties, as well as electrochromics-based and thermochromic coatings which enable a dynamic control of the throughput of radiant energy. The fascinating possibilities with electrochromic "smart windows" are pointed out. Angular-selective coatings are discussed briefly. Novel antireflection coatings make it possible to boost the transmittance both of coated and uncoated glass. Regarding materials interposed between the panes, the discussion includes gases and gas mixtures, coated plastic foils, silica aerogels, photochromic plastics, thermochromic cloud gels, and electrically switched liquid-crystal-based materials.

I. INTRODUCTION

Space conditioning of residential and commercial buildings accounts for substantial parts of the annual energy consumption in many countries. For the case of Sweden, this amounts to about 40 %. On the order of one sixth is required to offset heat losses through windows, i.e., roughly 7 % of the annual energy consumption is tied to fenestration performance. It is believed that similar numbers are valid for many other countries. Hence it is obvious that window properties have a significant effect not only on visual and thermal comfort but also on energy consumption at a global level. The purpose of this paper is to review current research and development on materials for energy efficient windows and to point out options in present and forthcoming technology. We will show how superior thermal insulation can be combined with good visual transmittance, how overheating can be avoided by reflecting off the infrared part of the solar spectrum, and how today's research on materials for optical modulation may lead to tomorrow's smart windows which will be able to regulate the throughput of radiant energy in accordance with dynamic needs. Earlier reviews on materials for energy-efficient windows have been given recently by Granqvist[1-5] and Lampert.[6,7]

We consider only windows composed of two or more panes of a transparent material (normally glass) with an interposed substance (normally a gas). The function of the window, in general terms, is to transmit a controlled amount of luminous radiation (for vision) and solar radiation (for space heating) at a specific - usually minimized - heat transfer. The heat transfer comprises additive contributions from thermal radiation, conduction in solids and gases, and gas convection. The ubiquitous radiation is of particular significance for at least four reasons: (i) the most fundamental object of a window is to transmit light, (ii) radiative transfer often accounts for a large fraction of the heat loss, (iii) radiative properties can be conveniently modified by thin coatings and surface treatments on the glass, and (iv) dynamic control of the radiative properties can be achieved in smart windows. Section II below lays the ground for a scientific discussion of energy efficiency by examining the radiation in our surroundings. It will be shown that luminous, solar and thermal radiation are confined to specific and well defined wavelength intervals. A key concept for energy efficiency is spectral selectivity, implying that the radiative properties should be qualitatively different for different wavelength ranges so that, for example, it is possible to combine transmittance of luminous radiation with reflectance (i.e., suppressed emission) of thermal radiation. Section II also defines integrated quantities which govern the radiative performance, and defines the goals one should aim for by proper materials selection. Section III contains some notes on the optical properties of standard window glasses. This information provides a baseline for subsequent discussions of ways to improve the energy efficiency. We also include some data for photochromic glass. Section IV treats a variety of means for using thin surface coatings to improve the radiative properties of glass. After a primer on large-scale coating technologies, we give a superficial discussion of various materials options. Then we follow with more elaborate accounts of spectrally selective noble-metal based coatings, spectrally selective doped oxide semiconductor coatings, electrochromic coatings with electrically controlled transmittance, thermochromic coatings with temperature dependent transmittance, angular-selective coatings, and certain novel antireflection treatments. So far we have only looked at ways to modify the radiative properties. However, conduction and convection are possible to control through the material interposed between the panes. These aspects are covered in Sec. V. The most radical control is achieved by introducing vacuum, in which case both conductive and convective heat transfer vanish. Less efficient, but much easier to accomplish in practice, is to work with gases which yield a lower heat conduction and a smaller convection than air. Convection can be prevented if the space between the panes is broken up into layers or cells with dimensions no larger than a few mm. If the cell sizes are much less than the wavelengths of visible light - which is the case of highly porous silica aerogel - the interposed solid material can be almost invisible. Section V also treats some means to accomplish variable transmittance by use of photochromic plastics, thermochromic cloud-gels, and liquid-crystal-based materials with electrically controlled opacity. Section VI, finally, gives a summary of the main results and attempts some predictions about the future of energy efficient architechtural windows.

It should be obvious from the above survey of the contents of this chapter, that there are a great many ways to improve the energy efficiency of windows. This paper does not purport to cover all of them in detail. Instead we focus on research and development which, in the author's view, holds particular promise for the future. Thus the presentation of the properties of standard window glass is brief, whereas the treatment of surface coatings - especially the electrochromic ones - is more elaborate. Economic estimates related to new fenestration technology are given only at a few places. The ongoing research on windows, and on the materials required for their energy efficiency, is very vigorous and carried out worldwide, and it seems premature to try to set down costs of products, such as the earlier mentioned smart window, which are in early stages of development and for which no generally agreed upon technology has been established.

II. ENERGY EFFICIENCY IN DIFFERENT CLIMATES

A. Ambient radiation

The key to energy efficiency of windows lies in a clear understanding of the radiative properties of our natural surroundings. These are illustrated in Fig. 1 with a common logarithmic wavelength scale on the abscissa. The most fundamental property is that matter emits radiation. Its character is conveniently discussed by starting from the ideal blackbody, whose emitted spectrum - known as the Planck spectrum - is uniquely defined if the absolute temperature is known. Planck's law is a consequence of the quantum nature of electromagnetic radiation. The right-hand part of Fig. 1 depicts Planck spectra for two temperatures of practical significance for windows. The vertical scale denotes power per unit area and wavelength increment (hence the queer-looking unit MWm^{-3}). The spectra are bell-shaped and confined to the $2 < \lambda < 50$ μm range, where λ signifies the wavelength. The peak in the spectrum for 50°C lies at a shorter wavelength than the peak in the spectrum for 0°C, which is a manifestation of Wien's displacement law. At room temperature the peak occurs at about 10 μm. Thermal radiation from a material is obtained by multiplying the Planck spectrum by a numerical factor - the emittance - which is less than unity. In general, the emittance is wavelength dependent.

The solid curve in the left-hand part of Fig. 1 reproduces a solar spectrum for radiation that has passed perpendicularly through the earth's atmosphere under typical clear weather conditions.[8,9] The curve has a bell shape corresponding to the sun's surface temperature (~6000°C). The integrated area under the curve is approximately 1000 Wm^{-2}; this is the largest possible power density on a surface oriented perpendicular to the sun in the presence of atmospheric damping. Solar radiation is seen to be confined to the $0.3 < \lambda < 3$ μm range. The minima in the solar spectrum are caused by atmospheric absorption, mainly by water vapour,

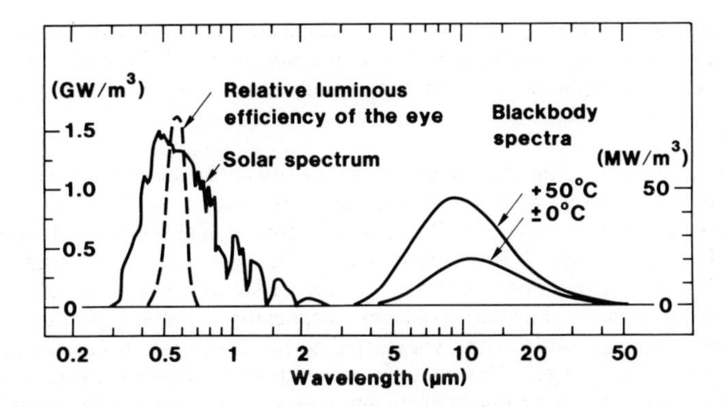

Fig. 1. Spectra for blackbody radiation at two temperatures, solar radiation that has passed through the earth's atmosphere, and relative sensitivity of the human eye.

carbon dioxide, and ozone. An interesting feature, which is not illustrated in Fig. 1, is that the transmittance through the atmosphere can be large in the $8 < \lambda < 13$ μm range, provided that the humidity is moderately low. Thus some thermal radiation can go almost unperturbed from an object at ground level into space, which gives a source of passive cooling with a power density of about 100 Wm^{-2} at ambient temperature[10] as we return to in the final chapter of this book. The fact that the sun subtends a small element of solid angle, whose position in the heavens is precisely known for each geographical location and for each time, opens possibilities to exploit surfaces with strongly angular-dependent properties.

The dashed curve in the left-hand part of Fig. 1 shows the relative spectral sensitivity of the human eye in its light-adapted (photopic) state. The bell-shaped curve extends across the $0.4 < \lambda < 0.7$ μm interval with its peak at 0.555 μm.[11] In its darkness-adapted (scotopic state), the eye's sensitivity is displaced about 0.05 μm towards shorter wavelengths. Photosynthesis operates with wavelengths in the same range as those for the human eye, which is of obvious relevance for greenhouse applications.[12,13]

B. What is Energy Efficiency?

An energy efficient window is a device capable of providing good lighting during the day and good thermal comfort both during day and night at mimimum demand of paid energy. Thermal comfort implies that overheating as well as excessive cooling should be avoided, that draught should be small, etc. Thus energy efficiency involves control of radiative inflow, and of heat losses due to radiation, conduction and convection. The radiative component to energy efficiency is conveniently discussed with reference to the above mentioned ambient radiation. It is suitable to make a separation into the requirements imposed by a warm, a cold, and a temperate climate.

In a *warm* climate it is frequently the case that the solar energy which enters through the windows and is absorbed in the room causes overheating. Space conditioning then requires air cooling equipment. It is clearly energy effective to have "solar control" windows which block the infrared part of the spectrum ($0.7 < \lambda < 3$ µm) without excessive lowering of the luminous transmittance ($0.4 < \lambda < 0.7$ µm). From the spectral distribution of solar energy it is inferred that, in principle, it is possible to *exclude about half of the solar energy at no decrease in luminous transmittance.* Another approach regards angularly dependent transmittance and is based on the fact that the luminous transmittance through windows usually has to be large only for near-horizontal lines-of-sight, whereas the sun is far above the horizon during most of the time when overheating is a severe problem. Thus having windows with properly tailored angular dependence of the transmittance is conducive to energy efficiency. Angular selectivity is of importance for inclined windows as elaborated on below. A combination of spectral selectivity and strongly angular dependent transmittance clearly is the superior option.

In a *cold* climate a window frequently causes an undesired loss of energy, and hence space conditioning involves heating. It is obvious that energy-efficiency hinges on a decrease of the heat losses. Convection can be diminished by use of multiply-glazed windows incorporating one or more slabs of essentially still gas. The heat transfer can be further lowered by diminishing the thermal radiation. In order to develop a feeling for the order-of-magnitude improvement one can accomplish, we reproduce in Fig. 2 some results of a study by Rubin et al.[14] Analogous data have been reported by Karlsson et al.[15] Rubin et al.[14] investigated windows with one, two, or three panes separated by air gap(s) of 12.7 mm. The surfaces are designated by consecutive numbers, with the outside surface labeled l. One of the surfaces is assigned an emittance in the 0-85 % range, where the upper limit refers to normal glass, and the thermal conductance - i.e., the k-value - is computed. The data refer to an outside temperature of -18° C and a wind speed of 24 km h^{-1}. The upper two curves show that if the emittance of one of the surfaces in a single-pane window is lowered, there is a marginal drop in the k-value from its magnitude of ~ 6 $Wm^{-2} K^{-1}$ for normal glass. In the double-glazed unit, the k-value can drop from ~2.8 $Wm^{-2} K^{-1}$ for normal glass to ~ 1.4 $Wm^{-2} K^{-1}$ when the emittance of either of the surfaces facing the air gap is brought to zero. For triple glazing, the corresponding improvement is from 1.8 to 1.2 $Wm^{-2} K^{-1}$. It is important to note that all of these improvements deal with radiation at $\lambda > 3$ µm, i.e., outside the solar range. One finds that it is possible, in principle, to *improve the thermal insulation of a double-glazed window by about a factor two at no decrease in solar transmittance.*

In a *temperate* climate there is sometimes a need for preventing excessive solar heating, whereas at other times one wants to let in as much as possible to provide free heating. Thus one wants to have a window which enables a dynamic throughput of radiant energy. Using established technology, one can work with mechanical regulation with movable shades, blinds or (roller) curtains and many designs are known (see for example Refs. 16 and 17). However a superior solution to the regulation is to invoke chromogenic materials,[18] which offer the possibility of changing the inflow of luminous and/or solar radiation in accordance with demands which can vary over the day or season. Devices of this type are called smart windows; this concept, introduced a few years ago by Svensson and

Granqvist,[19] seems to have gained general acceptance as a generic term. The regulation can be user operated or automatic. Among the many conceivable possibilities we may imagine a system in which a thermometer senses the indoor air and activates a low-voltage electric pulse to the window which sets its solar transmittance to a desired level. Chromogenic fenestration is readily combined with good thermal insulation through multiple glazing and low emittance. One concludes, that energy efficiency is tied to *smart windows incorporating materials which enable the throughput of radiant energy to be controlled between widely separated limits.*

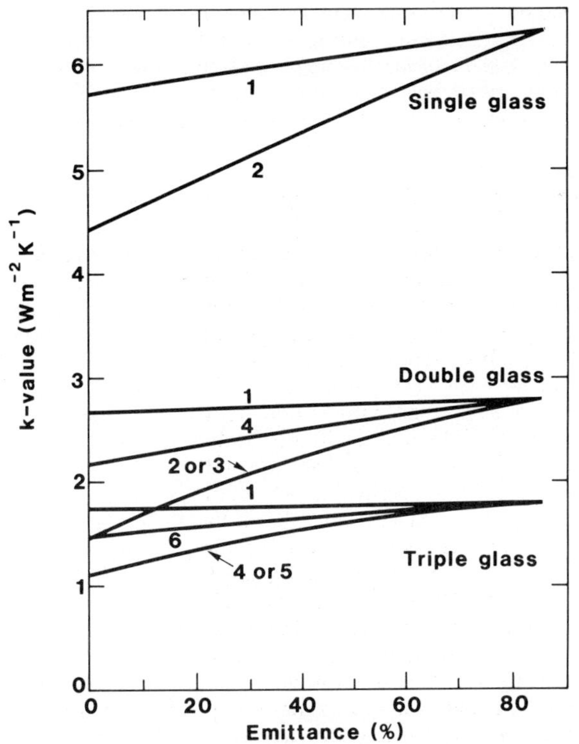

Fig. 2. Computed k-values of various window designs as detailed in the main text. (From Ref. 14).

Good thermal insulation improves energy efficiency also in a more indirect way than through a lowering of the k-value. To see this, one should note that if the insulation is good the surface temperature of the inner glass will remain close to that of the indoor air even if the outdoor air is very cold. This tends to significantly decrease the downwards stream of cold air along the surface of the window, which improves the comfort noticeably. It is important to observe that this draught is not connected with leaks at the window frame but is an inherent property of a cold gas. If draught is eliminated, the comfort temperture can be decreased by a few °C, space utilization can be more rational than if the area at the windows must be avoided, and heating installations can be simplified. All of these effects clearly contribute to energy efficiency. We point out, finally, that cold draught can be eliminated by electrically heated window glass. Commercial triple-glazed units allowing a peak load of 80 Wm^{-2} at 220 V exist on the market for linear sizes up to ~ 2 m. Whether or not they are energy-efficient depends on the design of the overall heating system.

C. Quantitative Performance Parameters for Energy-Efficient Windows

The idealized properties of energy-efficient windows, stated above, cannot be obtained with practical materials, and hence there is a need to specify suitable performance parameters which tell how far off a window is from certain design goals. The radiative properties comprise the absorptance A, reflectance R, transmittance T and emittance E, each signifying a certain fraction of incident, or maximum emitted, radiative power. Energy conservation dictates that at each wavelength one has

$$A(\lambda) + R(\lambda) + T(\lambda) = 1, \tag{1}$$

$$E(\lambda) = A(\lambda). \tag{2}$$

Equation (2), known as Kirchhoff's law, applies to matter in thermodynamic equilibrium.

Pertinent wavelength-averaged radiative properties can be defined by integrating over the eye's sensitivity curve, which gives the luminous (lum) performance, a typical solar (sol) spectrum, and a blackbody spectrum, which gives the thermal (therm) radiative performance. Quantitative data can be obtained from the general relation

$$X_\gamma(\theta) = \int d\lambda \; \phi_\gamma(\lambda) \; X(\lambda,\theta) / \int d\lambda \; \phi_\gamma(\lambda), \tag{3}$$

where X is A, R, T, or E; θ is the angle from the normal for the incident or emitted radiation; and γ denotes lum, sol or therm. In principle, angle-averaged properties can be specified by integration over θ.

For ϕ_{lum} it is proper to use the standard luminous efficiency function (Fig. 1), and for ϕ_{sol} one can use tabulated spectra appropriate to a certain air mass (AM).[9] For vertical windows it is often suitable to use the AM2 spectrum, corresponding to the sun being 30° above the horizon. For ϕ_{therm} one has

$$\phi_{therm} = c_1 \lambda^{-5} [\exp(c_2/\lambda\tau) - 1]^{-1}, \tag{4}$$

with $c_1 = 3.7418 \times 10^{-16}$ Wm^2, $c_2 = 1.4388 \times 10^{-2}$ mK, and τ signifying absolute temperature. In many cases it is convenient to specify the normal (θ=0) properties; to gain a simple notation we then omit the argument in the integrated optical property and write, for example, T_{lum} instead of $T_{lum}(0)$. However, the radiation of interest for thermal insulation takes place for all directions within a hemisphere (Siegel and Howell 1981), so that E_{therm} should be obtained from

$$E_{therm} = \int_0^{\pi/2} d\,(\sin^2\theta)\, E_{therm}\,(\theta). \qquad (5)$$

A distinctive feature of the ambient radiative properties is their spectral selectivity, i.e., their confinement to well-defined and sometimes non-overlapping wavelength intervals. Thus it is possible to have a window with a T_{sol}/T_{lum} ratio significantly less than unity, as desired for a warm climate, a large T_{sol} and a small E_{therm} as desired for a cold climate, and a variable T_{lum} or T_{sol} together with a small E_{therm}, as desired for a temperate climate.

In a non-evacuated double-glazed window, heat transport is connected with radiative transfer only to roughly 50 % and the full k-value is clearly a most significant parameter. It embraces effects of conduction and convection in the space between the panes. These latter aspects of heat transfer are not elaborated here, and we refer to the literature[21,22] for detailed treatises.

III. SOME NOTES ON THE TRANSMISSION THROUGH WINDOW GLASS

A. Standard Window Glass

The purpose of this section is to present a few selected optical data on standard window glass in order to give a baseline for subsequent discussions of means to improve the energy efficiency.

Normal windows are made by the float process in which the glass is solidified on a bath of molten tin. The uniformity and flatness of this glass are excellent. Figure 3 illustrates spectral transmittance in the solar range for three types of float glass. It is seen that T_{lum} is large. The transmittance in the infrared as well as in the ultraviolet are significant and dependent on the glass type. In the thermal infrared - not shown in Fig. 3 - glass is virtually opaque. The major difference among the glass types in Fig. 3 is their metal oxide content. With regard to energy efficiency, its most salient influence is to produce a broad absorption band centered at $\lambda \approx 1$ μm. If a maximum value of T_{sol} is desired, a low Fe_2O_3 content is preferrable; the upper curve in Fig. 3 refers to such a glass which has $T_{lum} \approx T_{sol} \approx 91$ %. The middle curve in Fig. 3 is valid for a normal type of float glass, which has $T_{lum} \approx 87$ % and $T_{sol} \approx 78$ %. The bottom curve, finally, shows the transmittance through a glass with a large metal oxide content. The absorption band extends somewhat into the visible and gives a greenish tint. This glass may be suitable for avoiding excessive solar heating. It is characterized by $T_{lum} \approx 72$ % and $T_{sol} \approx 47$ %. Thus the T_{sol}/T_{lum} ratio is 0.65, which seems to be about the lowest value one can reach in glass coloured by Fe_2O_3 or similar additives. Some of the absorbed energy will lead to a heating of the glass and a concomitant reemission of thermal energy, so that the total energy transmission is larger than T_{sol}, particularly for a tinted glass.

For the normal float glass according to the middle curve in Fig. 3, the total energy transmission (calculated with certain assumptions) is 83 %.

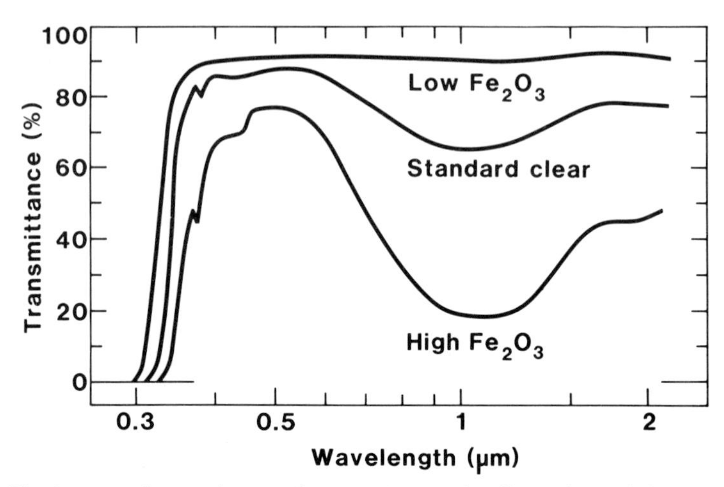

Fig. 3. Spectral normal transmittance for float glass of three different qualities. Specifically the curves refer to 6 mm Flachglas Optiwhite (upper), 6 mm Pilkington float glass (middle), and 4 mm Flachglas Flag Plus (lower).

We now consider the reflectance of the glass. In the spectral range where the absorption is weak, and at normal incidence, each air/glass interface has a reflectance governed by $(n-1)^2/(n+1)^2$, with n being the refractive index of the glass. In practice $n \approx 1.5$, so that each interface produces ~ 4 % reflectance. It is inferred that $T_{lum} < 92$ % is valid for a single pane. The overall transmittance is further diminished by multiple glazing. In the thermal infrared, the reflectance is low, which leads to a high emittance - in practice $E_{therm} \approx 85$ %.[23]

Laminated windows may be used for safety and other reasons. This glass comprises an interlayer of tough and resilient polyvinyl butyral (PVB) sandwiched between two glass panes and bonded under heat and pressure. The most salient optical effect of the PVB lies in the ultraviolet, which can be almost completely rejected as apparent from Fig. 4.

B. Photochromic glass

A photochromic material is characterized by its optical properties being able to change reversibly upon irradiation. The phenomenon is well known both in glasses[24,25] and other materials. A substantial photochromic effect in glass can be produced by adding special ingredients to the melt and by suitable melting and heat treatment procedures. In principle, it is possible to make use of isolated absorption centers in the vitreous matrix and photoelectronic processes (such as the reduction of Cd^{2+} to Cd^{+}), although it has as yet been difficult to prepare glasses which allow a sufficient number of colour/bleach cycles, i.e., which are fatigue-free.[26,27] Another, and more practical, approach to photochromic glasses rests on inhomogeneous microstructures with phases of photosensitivite compounds randomly dispersed in the vitreous matrix. Strong photochromic effects can be produced using metal halides - notably with silver and copper - as light absorbing substances.[24,25] Silver halide systems have reached the best technical maturity and are considered next.

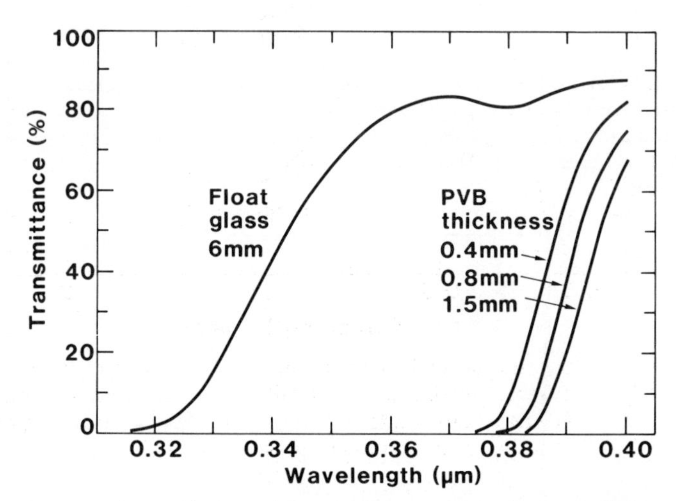

Fig. 4. Spectral normal transmittance through 6 mm float glass with and without a PVB laminate layer having the shown thickness. Reproduced from commercial information supplied by Monsanto Chemical Company, USA.

Photochromism based on silver halide particles is known in numerous optical glasses, for example in the alkali-alumo-boro-silicates, alkali-borates, lead-borates, lanthanum borates, and alumo-phosphates.[25] The alkali-alumo-boro-silicates are most widely used. These glasses are melted together with silver, chlorine and bromine ions added to the order of several tenths of a percent by mass. The amount of halogen ions exceeds the amount of silver ions. Cuprous ions to the order of 10^{-2} % by mass must be present in the glass melt. Essentially fatigue-free photochromism can evolve when such a glass is heat-treated above the glass transition temperature for a suitable time, which can be as long as several hours. Silver halide particles containing some Cu^{+} are then included in the glass matrix via a complicated phase separation process. Their diameter should be between 10 and 20 nm for maximum photochromism, minimum light scattering, and acceptable dynamics.[28]

For practical purposes, the optical properties of photochromic glass are governed by the darkened and cleared transmittance and by the darkening and clearing rates. These are dependent on the glass composition, and a cleared transmittance up to 90 %, or a darkened transmittance down to 5 %, are possible. The photochromism is somewhat temperature dependent, and a temperature rise yields enhanced transmittance and dynamics. Figure 5 shows T_{lum} during darkening and clearing for two different photochromic glasses at two temperatures. Irradiation is seen to bring down T_{lum} rapidly. Generally, darkening to ~80 % of the full range takes place in 1 minute, but a further minor darkening is measurable for times exceeding 1 h. Clearing in the absence of irradiation progresses slower than darkening and is incomplete even after 1 h. Figure 6 shows typical spectral transmittance in the $0.35 < \lambda < 1.3$ μm range for a photochromic glass in dark and clear states. It is inferred that the photochromic modulation of T_{sol} is significantly less than for T_{lum}.

The cost of photochromic glass is substantially higher than for standard glass owing to the manufacturing process (rolling, drawing, fusing) and the special heat treatment required. The float process has not yet been used for this process, as far as we know. At present (1991) photochromic glass is not produced for architectural applications.

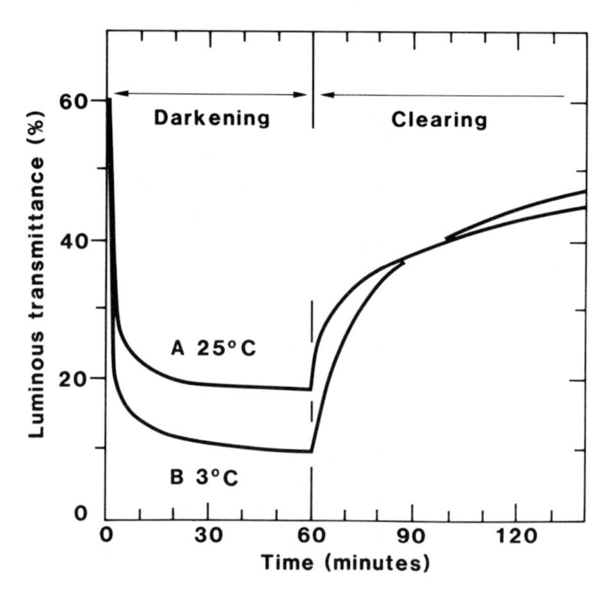

Fig. 5. Luminous transmittance vs. time during darkening and clearing of two photochromic glasses (denoted A and B). The glasses were cleared overnight prior to measurements. Data were taken at 20°C. Reproduced from commercial information supplied by Corning Glass Works, USA.

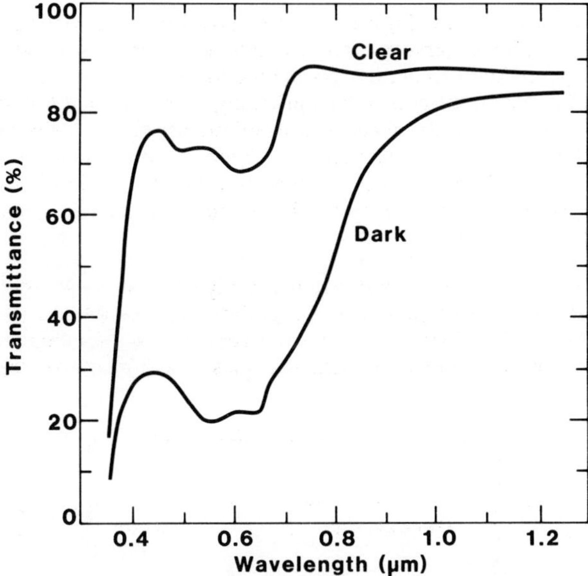

Fig. 6. Spectral normal transmittance for a photochromic glass in clear and dark states. Reproduced from commercial information supplied by Corning Glass Works, USA.

IV. COATINGS ON WINDOW GLASS

A. Coating Technology: A Primer

Surface coatings with thicknesses in the range 0.01 to 1 μm can modify the radiative surface properties of glass and thereby produce energy efficiency. Two techniques are in widespread use for preparing such coatings on the scale of square metres, viz. sputter deposition and spray pyrolysis. A detailed description of the technologies is not attempted here, but their operating principles and pros and cons will be outlined.

Figure 7a shows the principle of sputter deposition. The surface coating is prepared inside a vacuum chamber which contains an inert gas (usually argon) to a pressure on the order of one Pascal. The chamber holds one or more sputter cathodes whose lower parts comprise plates - known as targets - of the raw material for the coating. The glass passes in and out of the chamber by means of a load-lock system, and is transported a few cm below the targets. The deposition process involves a magnetically confined self-sustained plasma set up in such a way that energetic ions (usually Ar^+) bombard the target surface and dislodge atoms via complex momentum transfer processes. These atoms travel at high speed and stick to the glass, whose surface becomes uniformly coated. Use of direct current to power the plasma is customary and energy-efficient; it requires targets with some electrical conductivity. Radio frequency powering is an alternative for non-conducting targets. Dielectric thin films, for example of oxides, can be prepared by reactive sputtering in the presence of oxygen. A multilayer coating is conveniently produced by letting the glass pass under several cathodes which, if cross-contamination is feared, can be placed in separate chambers.

Figure 7b illustrates spray pyrolysis as a technique for making surface coatings. A solution, typically containing a metal chloride or acetylacetonate, is transported and dispersed through a system of nozzles by means of a carrier gas (air, nitrogen, argon, etc.) and, if required, a reactive gas. An aerosol is thus formed pneumatically and is sprayed towards the surface of a hot glass. The aerosol becomes vapourized before reaching the glass, and hence spray pyrolysis is a form of chemical vapour deposition. A typical reaction, of large significance for window coatings, is the hydrolysis of tin chloride to form tin oxide, shown schematically as $SnCl_4 + 2H_2O \rightarrow SnO_2 + 4\,HCl$.

(a) SPUTTER DEPOSITION

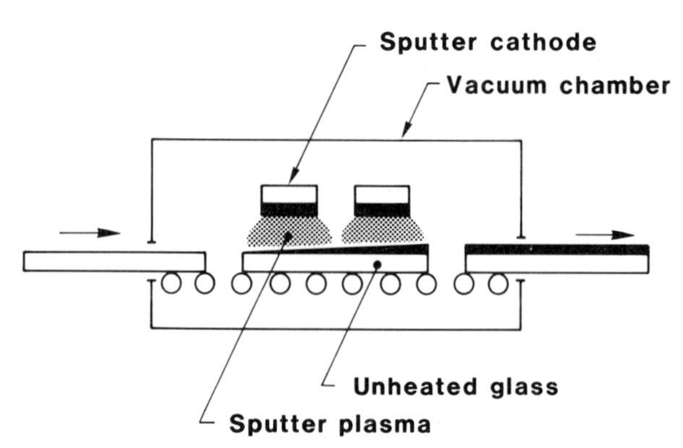

Fig. 7a.

(b) SPRAY PYROLYSIS

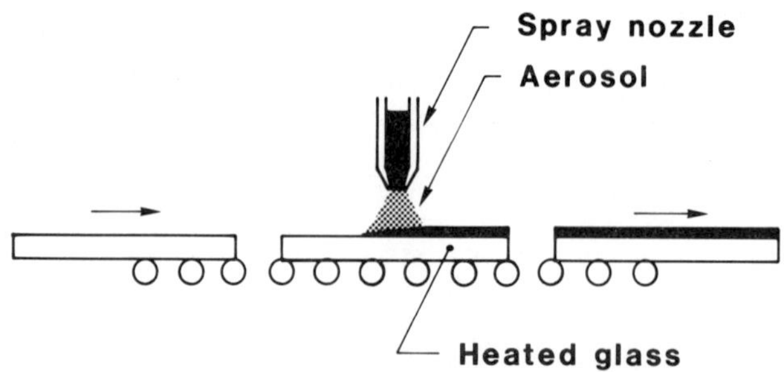

Fig. 7b. Principles for sputter deposition (a) and spray pyrolysis (b) as means to coat surfaces of glass transported as indicated by the horizontal arrows.

Sputter deposition as well as spray pyrolysis can be carried out by fully automatic equipment up to widths of several metres. Sputtering is notable for its versatility, possibilities to accomplish process control, multilayer facility, and low substrate heating (which makes it possible to coat plastic web and other temperature sensitive materials); on the negative side we note that high investment costs may be needed for equipment (of the order 10 MUSD for large-scale architectural coatings). Spray pyrolysis lends itself almost ideally to the production of extremely durable metal oxide based coatings by deposition onto the surface of a hot glass as it comes out from the tin bath of a float line. Multilayer deposition is possible.

Among the alternative practical window coating techniques there is notably dip coating, by which the glass is immersed in a chemical bath, withdrawn at a well-controlled rate (which governs the coating thickness), and heat treated. Further, vacuum evaporation is an old technique which can be used for example to metallize plastic web. For research on thin films, there are many other alternative coating techniques with specific advantages and disadvantages. Surface coating technology is a vast subject, which is covered to some depth in Refs. 29-33.

B. Objects and Materials Options

As a preamble to a discussion of the specific characteristics of different coatings for energy-efficient windows in subsequent sections, Table 1 summarizes the general goals of the coatings, the principle solutions to fulfill these, and the pertinent coating materials. Analogous tables for windows for automotive applications have been given recently.[34]

Diminished solar heating is desired for energy efficiency in a warm climate. About half of the solar energy comes as infrared radiation and can be excluded, in principle, with no effect on T_{lum}. Normal float glass has a significant transmission at $0.7 < \lambda < 3$ μm (cf. Fig. 3), and it is clearly effective to apply a surface coating whose reflectance is high preferentially at $\lambda > 0.7$ μm. Extremely thin continuous layers of the free-electron (here referred to as "noble") metals copper, gold, and silver can be used for this purpose. T_{lum} can be boosted by embedding the metal layer between high-refractive-index dielectric layers. A less efficient alternative is to start with a tinted glass - such as the one represented by the lower curve in Fig. 3 - and prevent thermal radiative inflow by a coating with low E_{therm}.

Diminished solar heating can also be obtained by combining high transmittance along a near-horizontal line-of-sight with a low transmittance for lines-of-sight which form large angles to the horizon. For vertical windows this calls for coatings whose transmittance falls off monotonically with increasing angle relative to the surface normal. For inclined windows, such as glass louvres, it is generally an advantage to have optical properties that are angularly selective. The meaning of angular selectivity is clarified in Fig. 8, which shows light beams incident at $\pm\theta_1$ and $\pm\theta_2$ to the surface normal of a plate such as a coated glass window. The transmittance values corresponding to $\pm\theta_1$ and $\pm\theta_2$ are denoted $T_{\pm 1}$ and $T_{\pm 2}$, respectively; the normal transmittance is denoted T_0. Angular selectivity refers to the property of having different transmittance for light of equal angle of incidence on either side of the normal. Thus a material with $T(+\theta) < T(-\theta)$ at a specific wavelength is said to have angular-selective transmittance at this wavelength. The angular selectivity can be tailored for different purposes: for example one can have a monotonic decrease in transmittance (e.g., $T_2 > T_1 > T_o > T_{-1} > T_{-2}$), or a uniformly high transmittance on one side of the normal and a uniformly low transmittance on the other side of the normal (e.g., $T_2 \approx T_1 >> T_{-1} \approx T_{-2}$). In practice, one can achieve angular selectivity with metal coatings having oblique columnar microstructures,[35-38] as will be discussed in Sec. IV G.

Table 1. General properties of coatings for energy-efficient windows.

Goal	Principle solution	Coating material*
Diminished solar heating	Reflectance at $0.7<\lambda<3$ µm	M or D/M/D
	Angular dependent transmittance	Oblique columnar metal
Thermal insulation	Reflectance at $3<\lambda<50$ µm	D/M/D, SnO_2:F; In_2O_3:Sn, ZnO:Al,....
Dynamic radiation control	Absorptance or reflectance in electrochromic material	Li_xWO_3, NiO_xH_y,.... in multilayer design with transparent ion conductor
	Reflectance at $0.7<\lambda<3$ µm in thermochromic material	VO_2-based
Higher transmittance	Antireflectance at $\lambda \approx 0.55$ µm	AlO_xF_y,....

*M is Ag, Cu, Au (or Al); D is Bi_2O_3, In_2O_3, SnO_2, TiO_2, ZnO or ZnS.

Good thermal insulation is imperative in a cold climate. Ordinary glass is characterized by a high thermal emittance, and it is effective to apply a surface coating with high reflectance at $3 < \lambda < 50$ µm. Two different types of coatings can be used to provide a low E_{therm}: an extremely thin metal film embedded between high-refractive-index dielectric layers with thicknesses chosen so as to maximize T_{sol}, and certain heavily doped oxide semiconductor layers. Among the latter we note SnO_2 doped by F or Sb, In_2O_3 doped by Sn, and ZnO doped by Al.

A dynamic throughput of radiant energy is highly desirable in a temperate climate, which calls for coatings of chromogenic materials. Two important approaches exploit materials exhibiting electrochromism[19,39] and thermochromism.[40] An *electrochromic* material is characterized by its optical properties being able to show a reversible and persistent change under the action of an electric field. The pertinent materials are transition metal (tungsten, nickel, etc) oxide layers, whose optical properties are altered by varying their content of small

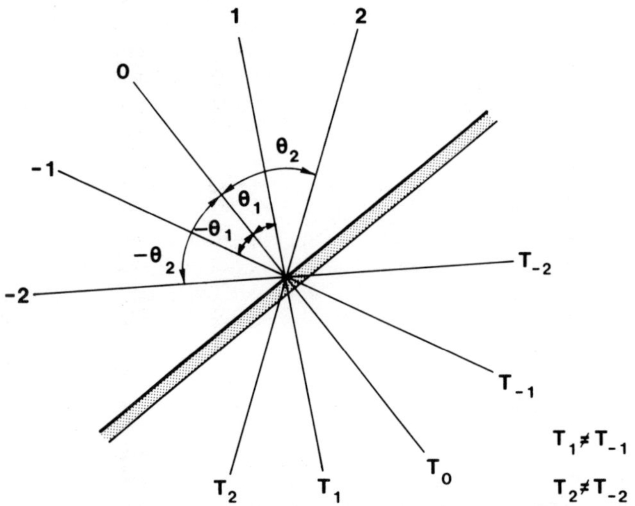

Fig. 8. Schematic illustrating the principle of angularly selective transmittance.

mobile ions (H^+, Li^+, etc). Additional layers serving as transparent ion storage, ion conductor, and electric conductors are required too, as will be discussed later. A *thermochromic* material has optical properties which depend reversibly on the temperature. VO_2-based coatings can have a transmittance which decreases upon a certain temperature being exceeded and are hence of interest for automatic radiation control. Surface coatings with good photochromic properties have been studied recently.[41] Pressure-dependent optical switching in SmS coatings - perhaps triggered by thermal expansion effects - also can be mentioned.[42,43]

Each air/glass interface yields a ~ 4 % reduction in T_{lum} and T_{sol}. Still larger reductions can be caused by surface coatings, particularly if they consist of materials with high refractive indices. A decrease of T_{sol} is undesirable in cold and, perhaps, temperate climates. A large R_{lum} may lead to visually disturbing effects of various kinds. The remedy for these problems is to apply an anti-reflection coating. A single layer designed for antireflecting glass (refractive index n) for visible light should have a refractive index of $n^{1/2}$ and a thickness of $\lambda/(4n^{1/2}) \approx 0.1$ μm. Certain novel high-rate-sputtered metal oxyfluoride coatings seem to have a great potential for antireflecting coated and uncoated window glass.[44-46]

For completeness we note that glass coatings can also be used to achieve glare and colour, thus producing interesting architechtural effects. As an example, Si coatings with high refractive indices can yield a durable reflecting glass. This kind of coating has almost no effect on E_{therm} and may, in fact, be detrimental to energy efficiency.

C. Noble-Metal Based Coatings

Very thin noble-metal coatings are, at least in principle, the simplest solution for reaching a significant short-wavelength transmittance combined with long-wavelength reflectance. The best optical properties are obtainable with copper, silver, and gold, as shown experimentally by Valkonen et al.[47] Alternative materials may be TiN (Refs. 48-51) and aluminium. Thin silver coatings stand out as the superior material on account of their low absorption of luminous and solar radiation.[52,53] The other metals are characterized by absorption, which lies at $\lambda < 0.5$ μm for copper, gold and TiN (hence their colour), and at $\lambda \approx 0.8$ μm for aluminium. Chemical and mechanical ruggedness is a critical issue for all noble-metal based window coatings, and results of laboratory experiments and field tests dictate that the coatings be used only on glass surfaces facing the gas enclosure(s) in hermetically sealed multiply-glazed units. The results of some durability tests have been reported recently.[54-56] Reviews of metal based window coatings have been given in Refs. 4, 57-59.

The limiting performance of a noble-metal based surface coating is readily computed, using established techniques,[60,61] by representing it as a plane-parallel slab of thickness t. The metal is characterized by parameters - known as the optical constants or the complex dielectric function - pertinent to a uniform and well-crystallized atomic arrangement.[62] Figure 9 shows computed results of T_{lum}, T_{sol}, R_{sol} and E_{therm} as a function of t for silver.[53] It appears that $t = 5$ nm yields $T_{lum} \approx 85$ %, $T_{sol} \approx 74$ % and $E_{therm} \approx 8$ %. If practically realizable, a glass with such a coating would be excellent for energy efficient windows. Unfortunately though, extremely thin noble-metal films can not be described as plane parallel slabs, as will be considered next.

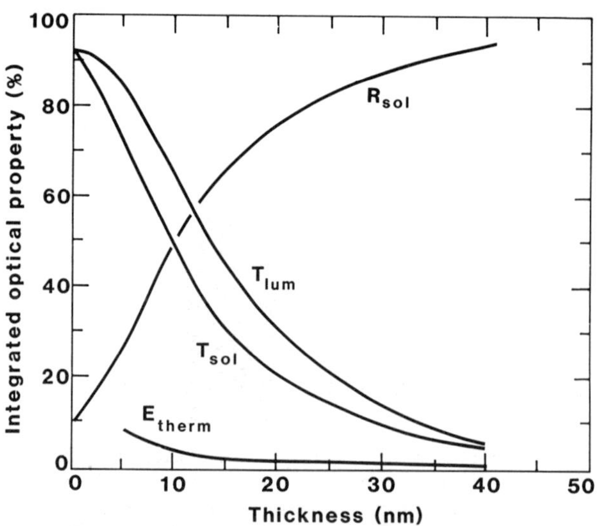

Fig. 9. Integrated optical properties versus thickness computed for an idealized plane-parallel silver slab. Replotted from Ref. 53.

When a metal such as copper, silver or gold is deposited onto glass, the coating goes through a series of rather well defined growth stages.[63] They are illustrated schematically in Fig. 10, where the pertinent structures and thickness scales are also shown. Initially, metallic nuclei will be formed at certain sites on the glass surface. Continued deposition makes the nuclei grow via surface diffusion and direct impingement. The metal islands thus formed undergo coalescence growth so that larger and more irregular islands appear. The growing coating then goes through large-scale coalescence so that an extended metallic network is formed. Subsequently the voids between the metallic paths become smaller and more regular. Finally, a uniform layer may be formed. The corresponding structures (with metal shown as black) are sketched. The size of the islands before large-scale coalescence is ~ 10 nm. Clearly, theoretical models based on a plane-parallel structure can be correct only for the later growth stages. Theoretical models of the optical properties are also known for thicknesses well below large-scale coalescence,[64,65] well above large scale coalescence,[66] and at the crossover.[67-71] Except at crossover, effective medium theories of the kind introduced in the chapter on Optical Properties of Inhomogeneous Two-Component Materials are of great value. The absolute thickness scale for the growth depends on many parameters such as the deposited species, the specific substrate material, the presence of (artificially added) nucleation centers, the substrate temperature, vacuum conditions, the presence of electric fields, and others. Figure 10 shows two typical scales referring to the deposition of gold onto glass by conventional evaporation (c.e.) and by evaporation in the presence of an optimized flux of energetic Ar^+ ions (i.e., ion-assisted evaporation, denoted i.a.) Deposition by sputter technology is believed to have an intermediate thickness scale. Currently there is a trend in sputter technology towards the use of "unbalanced" magnetron cathodes, which give a controlled ion bombardment of the substrate. The most important conclusion is that noble-metal films are reasonably uniform only at $t > 10$ nm.

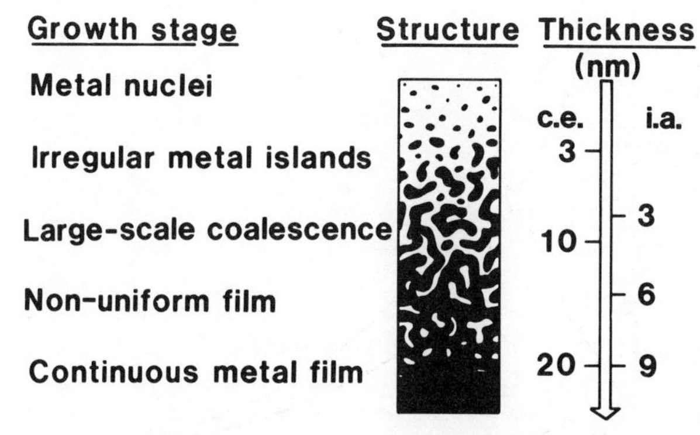

Fig. 10. Survey over growth stages, structures, and thickness scales for thin gold films deposited onto glass by use of conventional evaporation (c.e.) and ion-assisted evaporation (i.a.). (From Ref. 66).

It is now possible to appreciate the spectral optical properties as a function of t for films of copper, silver, gold, etc., deposited onto glass and similar materials. Figure 11, replotted from Valkonen et al.,[47] shows such data at $0.35 < \lambda < 16$ μm for conventionally evaporated silver layers. At t = 6 nm one finds that R_{therm} is low, and hence E_{therm} is high. This can be reconciled with the occurrence of an island structure; if the layer were continuous, E_{therm} instead would have been low as inferred from Fig. 9. It appears that a continuous silver layer is formed only at t > 9 nm. These latter coatings yield a high infrared reflectance combined with a significant T_{lum}, and hence they are of interest for energy efficient windows. Expectedly, T_{lum} and T_{sol} decrease with increasing t.

This tendency for thin noble-metal layers to form agglomerate structures at small thicknesses gives a significant limitation to their performance as window coatings. If high infrared reflectance is required, one is confined to $T_{lum} < 50\%$ and $T_{sol} < 40\ \%$ for layers produced by evaporation.[47] By optimized ion-assisted deposition[66] or special sputter technology[72] one can reach $T_{lum} \approx T_{sol} < 60\%$. If a still higher transmittance is desired one must use a multilayer coating.

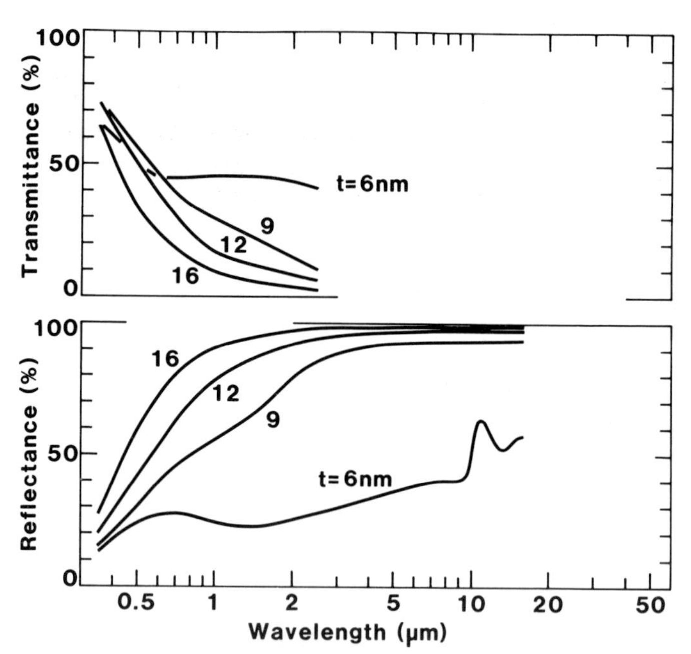

Fig. 11. Spectral normal transmittance and near-normal reflectance measured for glass coated with silver to the shown thicknesses by means of convectional evaporation. The reflectance peak at $\lambda \approx 11$ μm for the thinnest coating stems from the glass substrate. Replotted from Ref. 47.

The limited transmittance through reasonably uniform noble-metal layers is caused largely by reflection at the surfaces, and hence it is possible to improve the transmittance by additional layers which act so as to antireflect the metal. One is then led to consider dielectric/metal and dielectric/metal/dielectric multilayers. Dielectrics with high refractive indices - such as Bi_2O_3, In_2O_3, SnO_2, TiO_2, ZnO and ZnS - give the largest enhancement of the transmittance. By selecting the thicknesses of the three-layer configuration properly, one can optimize for a warm climate (maximum reflectance for $0.7 < \lambda < 3$ μm) or a cold climate (maximum T_{sol}; minimum E_{therm}). The optimization can be guided by computations which assume that the metal is a plane-parallel slab[15,53] or, more realistically, that the metal has thickness-dependent optical properties.[66]

Figure 12 depicts measured transmittance spectra, compiled from different sources, of coatings of the $TiO_2/Ag/TiO_2$ type on glass. These data are characteristic for what one can achieve by exploiting the three-layer design (see also Ref. 73). It is inferred that $T_{lum} > 80$ % and $E_{therm} << 20$ % are valid for all

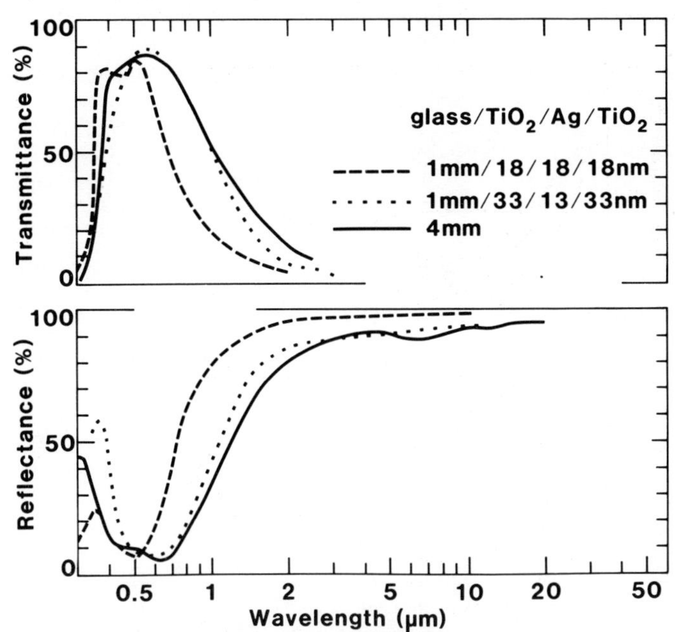

Fig. 12. Spectral normal transmittance and near-normal reflectance measured for $TiO_2/Ag/TiO_2$ based coatings on glass. Pertinent thicknesses for the glass and for the constituent layers of the coatings are shown. The solid curve stems from a commercially produced sample (Glaverbel Superlight), the dotted curve from Jarvinen (Ref. 74), and the dashed curve from Fan and Bachner (Ref. 52). The latter coating was originally devised to improve the luminous efficiency of incandescent light sources, but it is equally applicable as a window coating in a warm climate.

coatings. The dashed curve pertains to a coating with maximum infrared reflectance; it yields $T_{sol} \approx 50$ % and $R_{sol} \approx 42$ %. The dotted curve corresponds to $T_{sol} \approx 67$ % and $R_{sol} \approx 26$ %. Clearly, it is feasible to construct coatings for different climates simply by altering the thicknesses. The solid curve was measured for an optimized commercial coating on float glass of the type represented by the middle curve in Fig. 3. This latter coating yields $T_{lum} \approx 87$ % and $T_{sol} \approx 72$ %. Coatings of the type $TiO_2/TiN/TiO_2$ have been studied in Refs. 50, 51, 56. Noble-metal based three-layer coatings can be produced, using large-scale automatic sputter equipment, at a cost of about 10 USD/m^2 in the case of $TiO_2/Ag/TiO_2$, provided that target utilization is optimized.

D. Doped Oxide Semiconductor Coatings

Doped oxide semiconductor coatings offer an alternative to the earlier discussed noble-metal based coatings. The two classes of coatings have specific advantages and disadvantages. The semiconductor must be characterized by a wide bandgap, so that is allows good transmission in the luminous and solar ranges. Further it must allow doping to a level which makes the material metallic and hence infrared reflecting and electrically conducting. The materials which are known to be useful are all oxides based on zinc, cadmium, indium, tin, thallium and lead and alloys of these. The required doping is often achieved by the addition of a foreign element; particularly good properties have been obtained with SnO_2:F, SnO_2:Sb, In_2O_3:Sn and ZnO:Al. Another possibility is to provide doping via a moderate oxygen deficiency. If prepared properly, the mentioned coatings can be virtually non-absorbing for luminous and solar radiation. A specific and important advantage of the doped oxide semiconductors is their excellent chemical and mechanical durability, which allows their use on glass surfaces exposed to the air. As an extreme example, we note that pyrolytic SnO_2:F coatings have been successfully applied to the front side of "antifrost" windscreens on cars.[75] Reviews on doped oxide semiconductor coatings and their uses as transparent infrared reflectors and transparent conductors have been given in Refs. 4, 57-59, 76-81.

Figure 13 illustrates the principles by which the metallic properties are reached in at least most of the doped oxide semiconductors. The undoped crystal (for example In_2O_3) comprises a uniform arrangement of oxygen atoms and metal atoms. When dopant atoms (tin) are added, they replace some of the indium. If their density is sufficient, each dopant atom can be singly ionized by giving off an electron which - together with other liberated electrons - makes the material metallic. The onset of metallic properties occurs at a doping level of $\sim 10^{19}$ cm^{-3}. Electron densities up to $\sim 10^{21}$ cm^{-3} can be achieved by maximum doping, which implies that a few percent of the metal atoms in the oxide have been substituted by dopant atoms. For comparison, metals such as Ag and Au have an electron density of $\sim 6 \times 10^{22}$ cm^{-3}. Attempts to increase the doping still further leads to the formation of different absorbing complexes which diminish the transmittance. The infrared reflectance and the electrical conductance are governed by the scattering which the electrons undergo. It has recently been shown by Hamberg and Granqvist,[81] Jin et al.[82,83] and Haitjema et al.[84] that for properly produced In_2O_3:Sn, ZnO:Al and SnO_2:F coatings the scattering is dominated by the ionized impurities. The same is true for non-stoichiometric tin oxide.[85] This scattering is an unavoidable result of the doping, and hence Fig. 13 provides a unified picture of the basic processes which determine the infrared-optical and electrical properties of heavily doped oxide semiconductors.

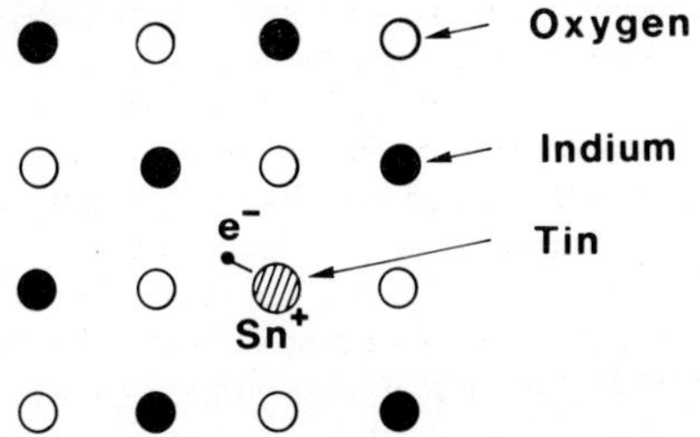

Fig. 13. Oversimplified crystal structure and doping model for In_2O_3:Sn. The actual crystal structure of In_2O_3 is rather complicated (cf. Ref. 86).

A detailed theoretical understanding of the doping mechanism makes it worthwhile to perform quantitative modelling of the spectral and integrated optical properties as a function of coating thickness and doping concentration (or, more precisely, electron density). Figures 14-16 contain data for In_2O_3:Sn coatings prepared by evaporation.[81] Disregarding fine details, the results can apply equally well to other deposition technologies and to alternative doped oxide semiconductor coatings.[58,82-84,87] Figure 14 shows spectral transmittance and reflectance for a 0.2-μm-thick coating whose electron density n_e lies between 10^{20} and 10^{21} cm^{-3}. The onset of infrared reflectance depends critically on n_e and it is seen that $n_e = 6 \times 10^{20}$ cm^{-3} yields high T_{sol} and high R_{therm} (i.e., low E_{therm}). High T_{lum} together with high reflectance at $\lambda > 0.7$ μm would demand that $n_e \approx 3 \times 10^{21}$ cm^{-3}, but such a doping level is inaccessible, as remarked above. One concludes that doped oxide semiconductor coatings are useful for energy efficient windows to be applied in a cold climate but not in a warm climate (unless an elaborate multilayer configuration is invoked; cf. Ref. 88).

Integrated optical properties are illustrated in Fig. 15. It is inferred that $T_{lum} \approx 80$ % is a typical value, that T_{sol} decreases significantly with increasing n_e and/or t, and that a low E_{therm} can be obtained only at $t > 0.2$ μm. The important relationship between T_{sol}, E_{therm} and t is further elaborated in Fig. 16. Requiring that $E_{therm} = 0.2$ results in a maximum T_{sol} of ~ 78 % at t = 0.2 μm. Demanding a lower E_{therm} leads to a significant drop in T_{sol}, while allowing a larger E_{therm} only gives a minor rise in T_{sol}. It is concluded that 0.2-μm-thick films with $4 < n_e < 6 \times 10^{20}$ cm^{-3} yield an optimum performance with $E_{therm} \approx 20$ % and $T_{sol} \approx 78$ %. Such properties can be achieved in coatings consisting of In_2O_3:Sn, ZnO:Al and SnO:F, and similar substances.

The required coating thickness equals a fraction of the wavelength for visible light, and it follows that optical interference effects must be given serious consideration. In particular, minor variations in t may cause significant iridescence, i.e., a rainbow exhibition of colours at different parts of the coated window. Such effects are not important for noble-metal based layers owing to

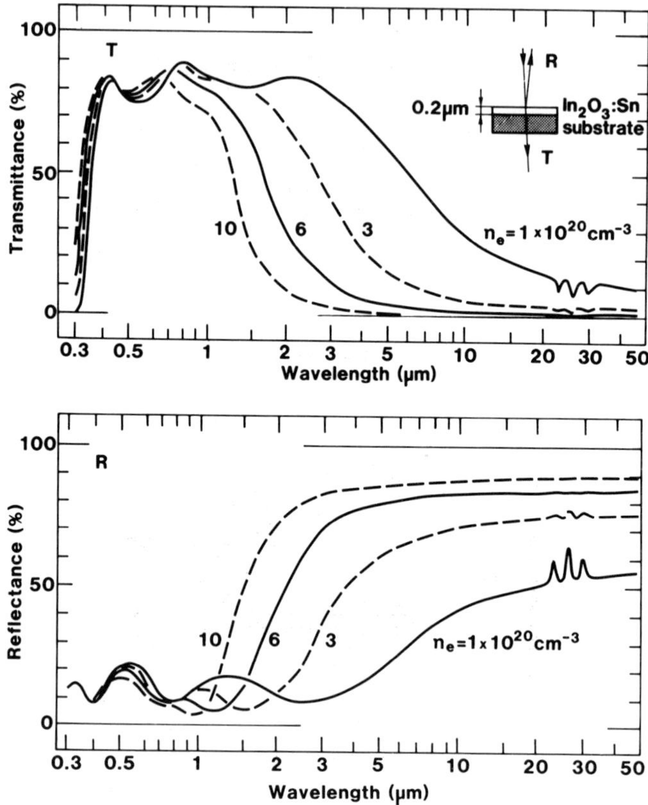

Fig. 14. Spectral normal transmittance (upper part) and reflectance (lower part) computed from a quantitative model for the optical properties of In_2O_3:Sn. The shown values of electron density (n_e) and coating thickness were used. (From Ref. 81).

their higher n_e's which allow $t << 0.2$ μm. Iridescent windows are generally considered aesthetically unpleasant, and it is concluded that extreme thickness control must be exercised during practical manufacturing of the coated glass. One possibility to avoid iridescence is to work with $t \sim 1$ μm. Such thick coatings yield numerous interference-induced peaks and dips of the transmittance and reflectance across the $0.4 < \lambda < 0.7$ μm interval. A thickness-dependent displacement of these peaks would not be recognized by the eye, which hence senses a uniform colouration. However, the use of a much larger thickness than the one demanded for a low E_{therm} is clearly inefficient in terms of materials utilization, coating time, and cost. Further, thick oxide semiconductor coatings can display some light scattering, sometimes referred to as haze. This phenomenon is associated with the occurrence of large crystallites as well as surface roughness.[89,90]

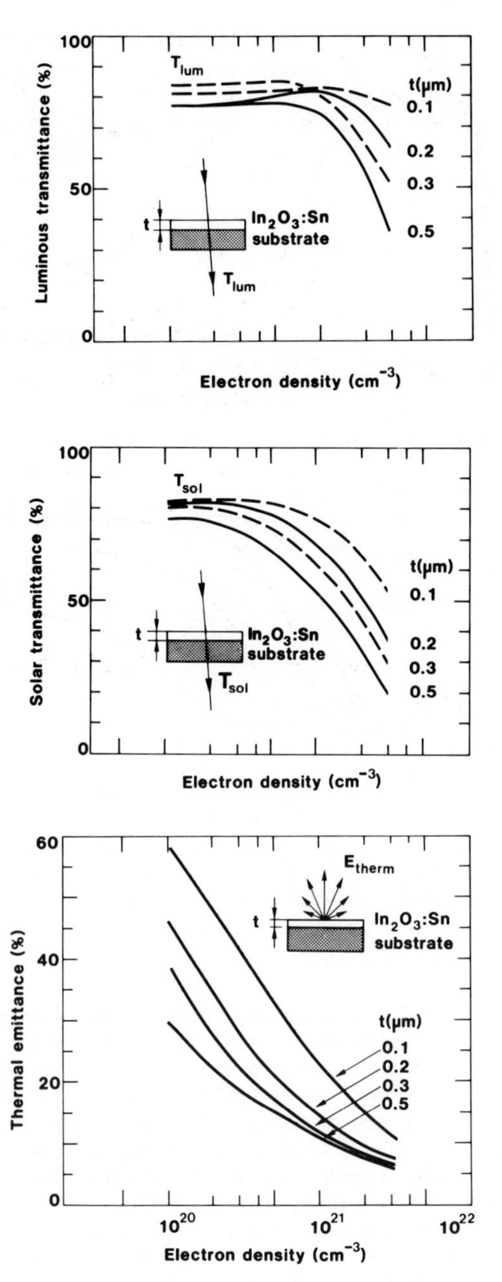

Luminous transmittance (%)
T_{lum}
t(μm)
0.1
0.2
0.3
0.5
In_2O_3:Sn
substrate
t
Electron density (cm^{-3})
Solar transmittance (%)
T_{sol}
Thermal emittance (%)
E_{therm}
10^{20}
10^{21}
10^{22}

Fig. 15. Luminous transmittance (upper part), solar transmittance (middle part) and thermal emittance (lower part) computed from a quantitative model for the optical properties of In_2O_3:Sn. The used values of electron density and coating thickness (t) are shown. The substrate was represented by properties pertinent to pure SiO_2 (i.e., a glass with no Fe_2O_3 content). (From Ref. 81).

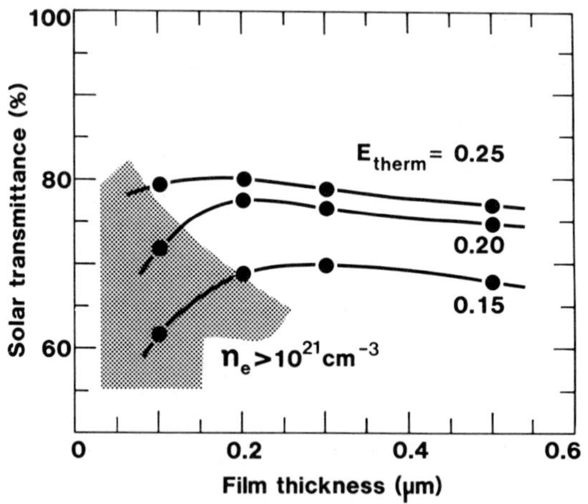

Fig. 16. Normal solar transmittance vs. coating thickness for three values of hemispherical thermal emittance computed for In_2O_3:Sn in the same way as in Figs. 14 and 15. The shaded area refers to hypothetical coatings with $n_e > 10^{21}$ cm^{-3}. (From Ref. 81).

Figure 17 shows transmittance and reflectance spectra as measured for three different doped oxide semiconductor coatings. The curves show rather high T_{lum}, moderately high T_{sol}, and indicate low E_{therm}. The upper part refers to a 0.41 μm thick In_2O_3:Sn coating prepared for research by evaporation technology.[81] This coating matches the theoretical predictions for an optimized material very well. The middle part concerns a 0.30 μm thick ZnO:Al coating prepared for research by sputter technology.[82,83] Both of these coatings were put on CaF_2 substrates that are transparent for $0.3 < \lambda < 10$μm; the data would have been very similar for coatings backed by glass. The bottom part was recorded for a SnO_2:F coating produced commerically by spray pyrolysis onto 4 mm standard floatglass. The latter coating is considerably thicker than the other two, as evidenced by the densely spaced peaks in the spectra. It has $T_{lum} \approx 74$ %, which is lower than for the In_2O_3:Sn and ZnO:Al coatings, and $T_{sol} \approx 77$ %. Commercial products with somewhat better data have become available recently. Some data on doped SnO_2 coatings have been summarized by Karlsson et al.[73]

Ultraviolet rejection is of importance for preventing bleaching of textiles, degradation of polymers, etc. Choice of proper glass (Fig. 3) as well as lamination (Fig. 4) are important for this purpose. ZnO coatings offer another possibility, as

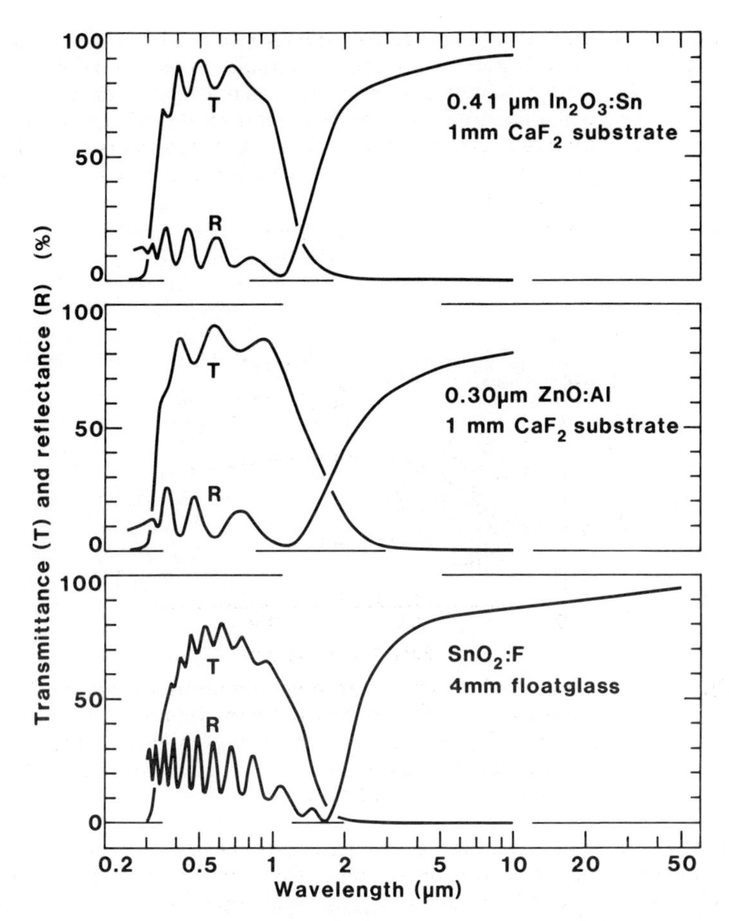

Fig. 17. Spectral normal transmittance and near-normal reflectance measured for In_2O_3:Sn on 1 mm CaF_2, ZnO:Al on 1 mm CaF_2, and SnO_2:F on 4 mm floatglass (Glaverbel Comfort Glass). Coating and substrate thicknesses are as given. Partly compiled from Refs. 81-83.

shown in Fig. 18. It appears that a 0.3 μm thick ZnO layer is sufficient for absorbing most of the ultraviolet radiation which is otherwise able to pass ordinary floatglass. Heavy doping, for example by Al, shifts the absorption to smaller wavelengths, and the ultraviolet absorption becomes insignificant. This phenomenon - known as a bandgap shift - is theoretically well understood.[91,92] Coatings based on In_2O_3 and SnO_2 are not useful for ultraviolet absorption. Tandem coatings of ZnO/ZnO:Al are able to combine large T_{lum} and T_{sol} with low E_{therm} and ultraviolet rejection. CeO_2 coatings may be applicable for the same purpose.[93,94]

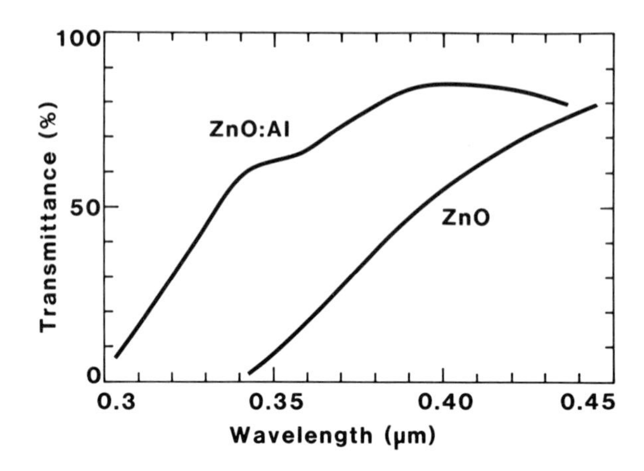

Fig. 18. Spectral normal transmittance in the ultraviolet measured for 0.3 μm thick ZnO and ZnO + 2 at.% Al coatings on 1 mm CaF_2.

The cost for large-scale doped oxide semiconductor coatings is dependent on production technology, as well as material, and is difficult to estimate. With regard to technology, it should be pointed out that several glass manufacturers have had success in their development of SnO_2-based coatings prepared by spray pyrolysis directly onto the hot glass on a float line. This technology holds promise for very inexpensive coatings, having costs of only a few USD/m^2. With regard to material, we note that indium is much more scarce and expensive than zinc and tin. Hence In_2O_3:Sn coatings should be used only when the ultimate in electrical conductivity or infrared reflectance are required. Cadmium and thallium based coatings may cause heath hazards.

E. Electrochromic Coatings

Electrochromism is a multi-facetted phenomenon which is well known in transition metal oxides based on tungsten, vanadium, nickel, molybdenum, titanium, iridium etc., and in numerous organic substances. The change in the optical properties is caused by the injection or extraction of mobile ions. A material that colours under injection (extraction) of ions is referred to as cathodic (anodic). Absorptance modulation as well as reflectance modulation are feasible. Electrochromism has been studied since the early 1970's for high contrast non-emissive display applications.[18,95-100] Since the mid-1980's, research and development on transparent electrochromic materials and devices for smart window applications have soared both in academia and in industry.[18,19,39,101]

An electrochromic coating must include several layers. Figure 19 sketches a basic five-layer configuration which is convenient for discussing the operating principles. It comprises two outer transparent electrical conductors, required for setting up a distributed electric field, an electrochromic layer, an ion conductor, and an ion storage. Colouration and bleaching are accomplished when ions are moved from the ion storage, via the ion conductor, into the electrochromic layer, or when the process is run in reverse. The ion storage can be another electrochromic layer, preferrably anodic if the base electrochromic layer is cathodic, or vice versa. One may also combine the ion conductor and storage media into one layer of electrolyte. Further, it is possible to exclude the ion storage and instead rely on a replenishment of H^+ ions (protons), originating from the dissociation of water molecules diffusing in from an ambience with controlled humidity; obviously this requires a substantial atomic permeability of the outer transparent conductor. By use of a purely ionic conductor, one can obtain an open circuit memory, i.e., the electric field has to be applied only when the optical properties are to be altered. The design in Fig. 19 has a low value of E_{therm} because of the external transparent conductor. Alternatively, the electrochromic coating can be combined with a lamination material which joins two glass panes;[102] in this case an additional transparent and infrared-reflecting coating is required to obtain a low E_{therm}.

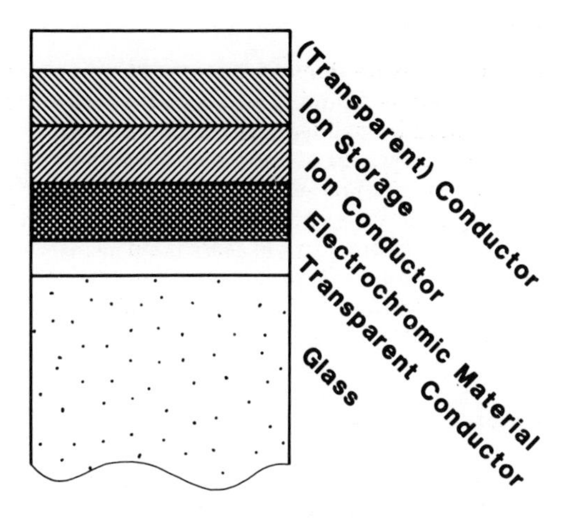

Fig. 19. Basic design of an electrochromic coating for smart windows. (From Ref. 19).

We now consider the materials in the different components of the electrochromics-based coating and first look at the actual electrochromic layer(s). Tungsten oxide was the first discovered electrochromic material and is by far the most widely studied one.[99] In amorphous state it permits the absorptance to be modulated between wide limits. In crystalline state it allows a fair degree of reflectance modulation.[103-105] The degree of crystallinity depends on the glass temperature during the coating process. Hydrated nickel oxide is a rather recently discovered electrochromic material which allows absorptance modulation.[106-120] Its durability seems to be superior to that of tungsten oxide. The mechanisms for colouration and bleaching of the mentioned electrochromic materials have been given as

$$xM^+ + xe^- + WO_3 \underset{\text{bleach}}{\overset{\text{colour}}{\rightleftarrows}} M_xWO_3$$

and

$$Ni(OH)_2 \underset{\text{bleach}}{\overset{\text{colour}}{\rightleftarrows}} NiOOH + H^+ + e^- ,$$

where M^+ denotes H^+, Li^+, Na^+, etc. Hence tungsten oxide colours under ion insertion whereas hydrated nickel oxide colours under ion extraction. This opens the possibility of working with an optically efficient "complementary" design incorporating layers of both materials as "electrochromic material" and "ion storage", respectively. It is also technically possible to operate tungsten oxide and iridium oxide in conjunction.[121]

The ion conductor can also be of many materials.[18] Disregarding liquid electrolytes - which are not practical for windows - it is possible to employ LiF, $LiAlF_4$, $LiNbO_3$ and the like[105,122] as well as layers of transparent polymeric ion conductors.[102,121,123,124] In devices which rely on ambient humidity, layers of ion-permeable SiO and MgF_2 coatings have been used successfully.[99,125-128] The transparent conductor on the glass surface can be of In_2O_3:Sn or a similar doped oxide semiconductor. The outer transparent electrode can consist of the same material or - if atomic permeation is required - of an extremely thin gold layer. The voltage required for the distributed electric field is on the order of a few volts and needs only be applied when the optical properties are to be altered.

Figure 20 illustrates the characteristic features of absorptance-modulated electrochromic coatings based on amorphous tungsten oxide and hydrated nickel oxide in fully bleached state, fully coloured state, and at intermediate colouration. The upper part refers to a device with two In_2O_3:Sn-coated glass substrates, one of them being overcoated with tungsten oxide, and an intervening liquid Li^+-containing electrolyte.[129] The lower part was obtained for one In_2O_3:Sn-coated glass with a hydrated nickel oxide overlayer made by radio frequency sputtering; the colouring was accomplished in a potassium hydroxide electrolyte prior to the optical measurements.[112] Both devices show pronounced electrochromism with the tungsten oxide absorbing especially strongly in the infrared and the hydrated nickel oxide absorbing especially strongly for visible light. No data are shown for $\lambda > 1.2$ μm owing to the influence of the In_2O_3:Sn layers (cf. upper part of Fig. 17).

Figure 21 shows quantitative results for the modulation of T_{lum} and T_{sol} by a glass plate with low Fe_2O_3 content overcoated with moderately conducting In_2O_3:Sn and with electrochromic hydrated nickel oxide prepared by radio frequency sputtering. It is seen that T_{lum} can vary between 80 and 10 % and that T_{sol} can

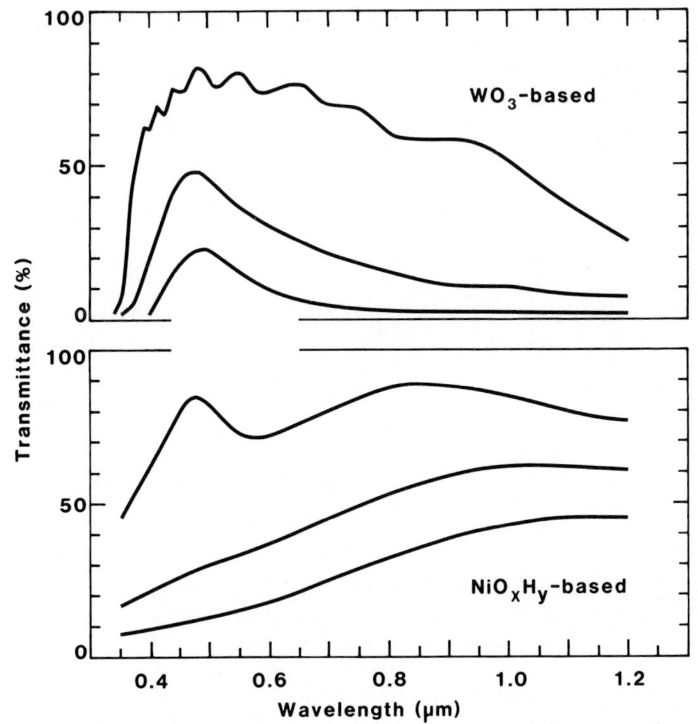

Fig. 20. Spectral normal transmittance mesured for research-type devices incorporating an electrochromic layer of tungsten oxide (upper part) and hydrated nickel oxide (lower part). Compiled from Refs. 112 and 129.

vary between 80 and 20 % when a charge density up to 20 mC/cm^2 is injected. Analogous electrochromic coatings made by direct current sputtering have not yet given results that are quite as good.[116]

In principle, reflectance modulation is superior to absorptance modulation, since the coated glass does not become heated to a comparable extent. Figure 22 shows reflectance spectra for a 0.23-µm-thick crystalline WO_3 layer on In_2O_3:Sn-coated glass operated in a Li^+-containing liquid electrolyte.[105] Li^+ injection is seen to give reflection, particularly in the infrared, up to ~70 % at λ = 2.5 µm. Extraction of the Li^+ ions brings back the initial low reflectance. Theoretical modelling of the electrochromic reflectance modulation has been performed by using the theory for the optical properties of doped oxide semiconductors.[81,127] The limiting optical performance - obtained with several simplifying assumptions - is that modulation can be accomplished between a state with $T_{sol} \approx 82$ %, $T_{lum} \approx 74$ % and $R_{lum} \approx 26$ % and another state with $T_{sol} \approx 35$ %, $T_{lum} \approx 63$ % and $R_{lum} \approx 5$ %. The theoretical understanding of absorptance-modulated electrochromic materials is yet too fragmentary to permit quantitative modelling.

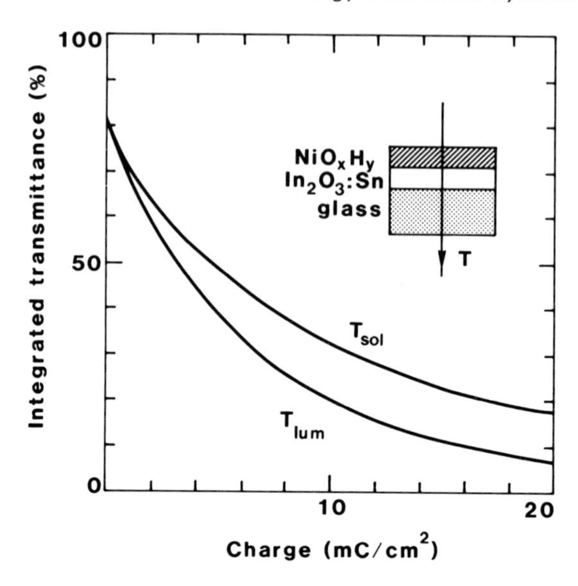

Fig. 21. Integrated transmittance vs. charge density for electrochromic coatings of the type shown in the inset. (From Ref. 112).

In the remainder of this section on electrochromics-based coatings, solid-state configurations are focused on, since these are the only ones of concern for practical window applications. Figure 23 shows transmittance spectra for a multilayer design with 0.15 μm of WO_3, ~ 0.1 μm of MgF_2, and 0.015 μm of gold. This type of device was originally developed by Deb[99] and has since been worked on by others. It relies on ambient water, which is catalytically decomposed at the Au electrode and driven via the MgF_2 layer into the amorphous electrochromic WO_3 layer by a voltage between the Au and In_2O_3:Sn electrodes. The data in Fig. 23 correspond to a modulation of T_{sol} between 25 and 3 %. It is not possible to obtain $T_{sol} >> 25$ % because of reflectance at the Au electrode. The devices become non-functional in a dry ambience but regain their electrochromism in a humid atmosphere.

The state-of-the-art (1991) in electrochromic smart windows is believed to be represented in Fig. 24. The investigated device incorporates two glasses with low Fe_2O_3 content, each having a transparent conducting layer and an electrochromic layer, laminated together by a solid transparent polymeric ion conductor. One of the electrochromic layers is anodic and the other is cathodic. The device is mechanically rugged and permits extended colour-bleach cycling. Absorptance modulation yields T_{lum} between 68 and 9.5 % and T_{sol} between 63.5 and 7.5 %. A low E_{therm} can be obtained by having a layer of SnO_2:F, for example, on either of the exterior glass surfaces.

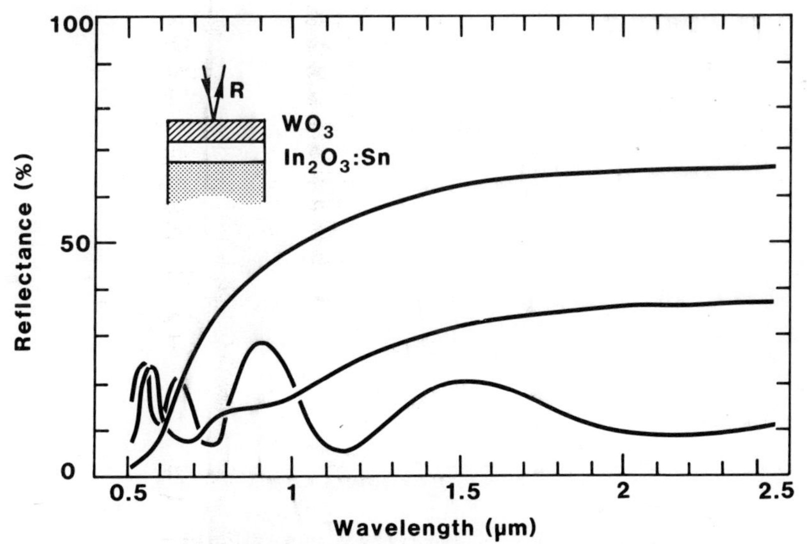

Fig. 22. Spectral near-normal reflectance measured for the electrochromic device shown in the inset at maximum infrared reflectance. (From Ref. 105).

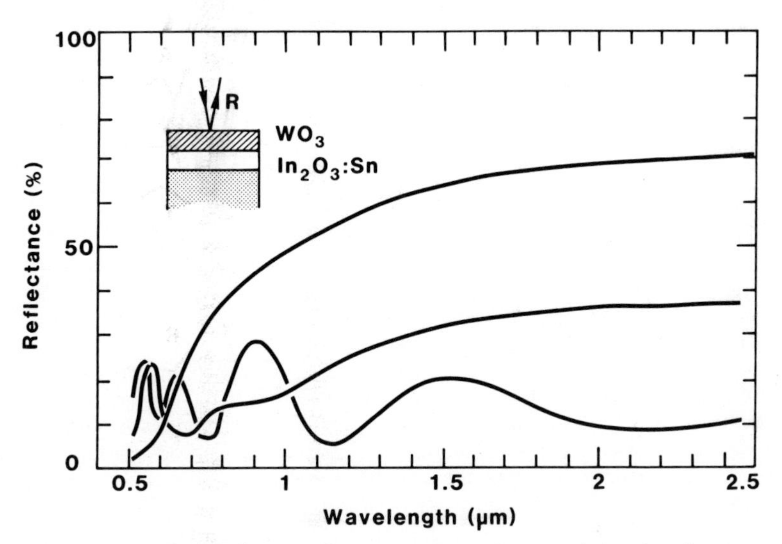

Fig. 23. Spectral normal transmittance measured for the electrochromic device shown in the inset in fully bleached state, in fully coloured state, and at intermediate colouration. Replotted from Ref. 19.

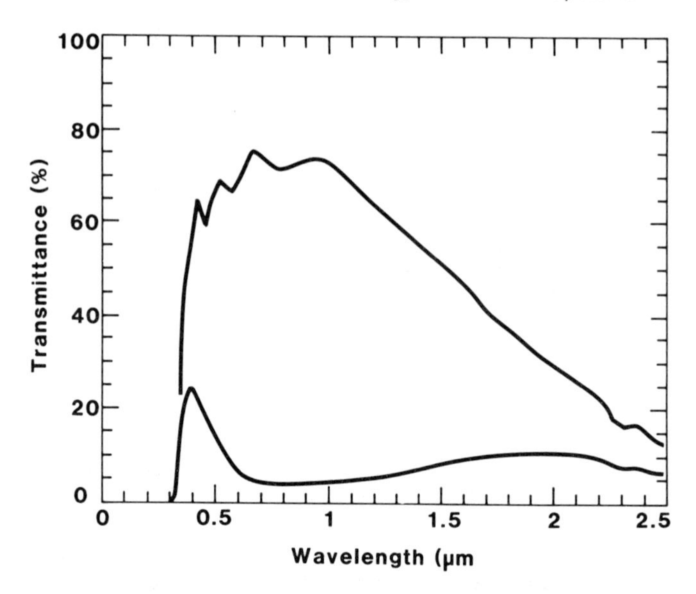

Fig. 24. Spectral normal transmittance measured for a research-type electrochromic smart window in fully bleached and fully coloured states.

This section is concluded with some practical considerations for smart windows. The time for going from fully bleached to fully coloured state, here called the response time r, is of obvious significance. The required time scale is set by the eye's ability to accomodate, which is on the order of minutes. We note, in passing, that a small display element should have $r \ll 1$ s. Following an analysis by Viennet et al.,[130] a 1 m^2 WO_3-based electrochromic layer can have a response time on the order of one minute. The magnitude of r is critically dependent on the electrical resistance in the transparent conductor(s); the lowest resistivity for In_2O_3:Sn is ~ 10^{-4} Ω cm.[81,131] The temperature dependence of r is elucidated in Fig. 25 with regard to the bleaching of a small WO_3-based sample.[132] It is seen that r rises at low temperatures, but the effect is of little practical significance for an electrochromic coating placed on the inner glass of a double-glazed window. Full colour uniformity over an extended surface cannot be expected during changes of T_{lum} and T_{sol}, but the uneveness can be made invisible at the expense of a long response time. Iridescence may be a concern, just as for doped oxide semiconductor coatings, for electrochromic multilayer coatings placed on a single glass surface. Electrochromic coatings operated in a laminate configuration are expected to be less prone to show iridescence owing to the matching of refractive indices for adjacent layers.

It is not possible to give even a coarse estimate of the cost of a practical smart window. However, it should be noticed that inherently cheap techniques are being developed for efficient deposition of electrochromic layers by chemical

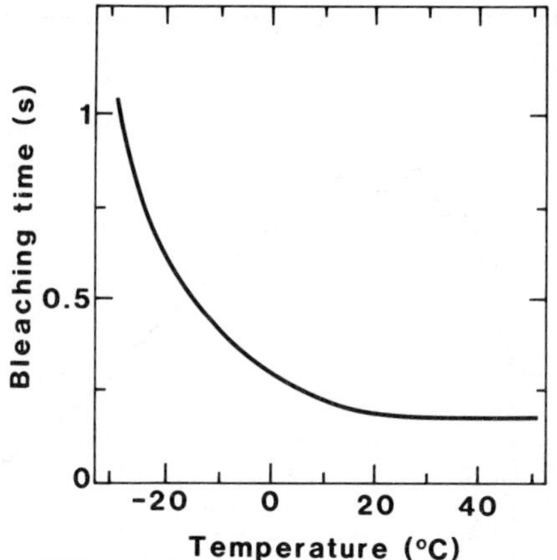

Fig. 25. Bleaching time vs. temperature for an electrochromic device with a WO_3 layer. (From Ref. 132).

vapour deposition[128,133,134] and dip coating.[135,136] Plasma enhanced chemical vapour deposition and high-rate sputtering are other technologies that may be used for practical large-scale production.[18] The electrical power needed for operating an electrochromic smart window can be assessed from the relation between charge density and colouration (cf. Fig. 21). A simple analysis yields that the electrical energy requirement is as low as $\ll 1$ kWh per year and square metre window area.

F. Thermochromic Coatings

Thermochromism is well known for inorganic materials in the liquid and solid state as well as in many organic materials.[18,137] Among the many thermochromic transition metal oxides,[138] vanadium dioxide stands out as particularly interesting for applications on smart windows.[40,139-142] VO_2 undergoes a structural transformation at a "critical" temperature τ_c, below which the material is semiconducting and relatively non-absorbing in the infrared, and above which the material is metallic and infrared reflecting. A window with a VO_2 coating hence has a T_{sol} which drops at $\tau > \tau_c$, i.e., the window is capable of an automatic control of the throughput of radiant energy. This control may be achieved by use of a single layer, i.e., with a coating design which is simpler than the one in an electrochromic smart window. However, the control is built into the material and is not easily modified by an operator.

Crystals of VO_2 are characterized by $\tau_c \approx 68°$ C, which is too high for normal window applications, but τ_c can be depressed by several techniques[40,142] such as by replacement of some vanadium by tungsten, molybdenum, niobium or rhenium, by replacement of some oxygen by fluorine, or by introducing stress either by use of a suitable substrate or by applying an overlayer. For the latter option, the overlayer can serve also so as to antireflect the VO_2 coating.

Figure 26 illustrates the thermochromism one can obtain in a 0.13-µm-thick vanadium oxyfluoride layer prepared by sputter deposition onto glass with low Fe_2O_3 content.[143] The coating is characterized by $\tau_c \approx 52$ °C (as obtained from electrical conductance), $T_{lum} \approx 28$ % irrespective of temperature, and T_{sol} ranging from 35 % at 25 °C to 28 % at 70 °C. T_{lum} is increased by the fluorination,[144] as apparent from Fig. 27; however the limited magnitude of T_{lum} may still be an obstacle for many window applications. The situation can be improved by use of an overlayer, though. Figure 28 shows spectral transmittance at $\tau << \tau_c$ and $\tau >> \tau_c$ for a glass coated by 0.05 µm of VO_2, made by evaporation, and 0.11 µm of SnO_2.[142] The double-layer has $\tau_c \approx 49$ °C , $T_{lum} \approx 45$ %, and $T_{sol} \approx 53$ % at 25 °C and $T_{sol} \approx 46$ % at 80 °C. The bare VO_2 layer has $T_{lum} \approx 35$ %.

Thermochromic VO_2-based coatings studied so far have an undesirably large A_{lum}, which tends to make the modulation of T_{sol} rather small. Nevertheless, thermochromic smart windows are of considerable potential interest, and basic materials research in this area is being pursued.

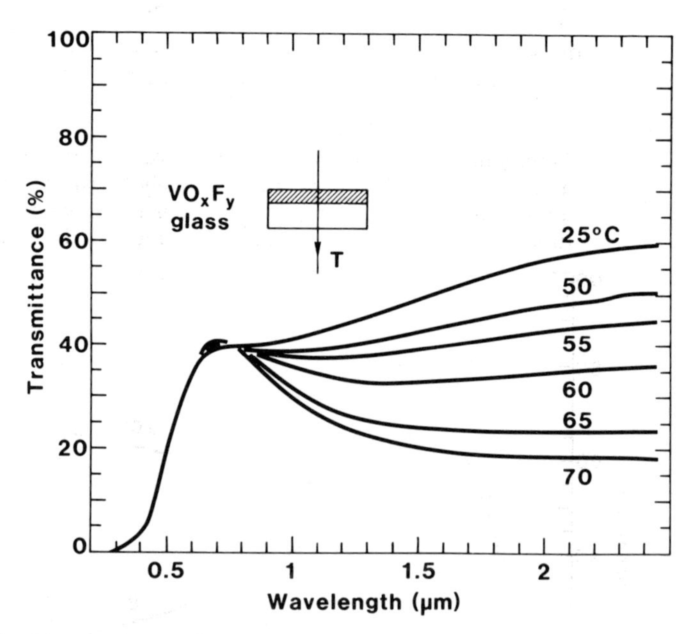

Fig. 26. Spectral normal transmittance measured for a research-type thermochromic smart window with a vanadium oxyfluoride coating at the shown six temperatures. (From Ref. 143).

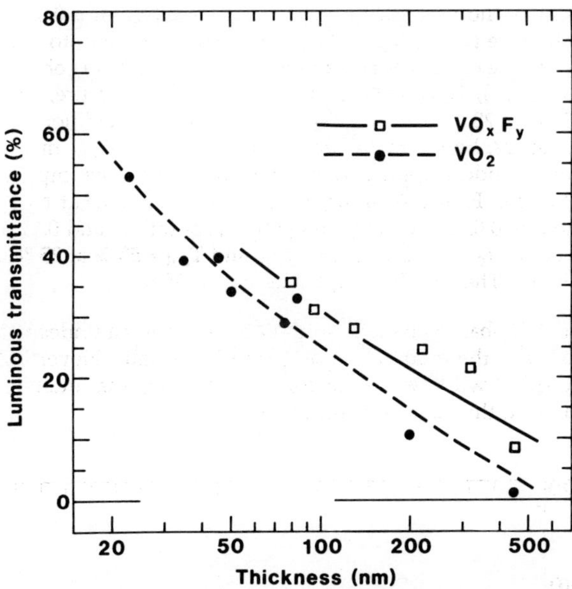

Fig. 27. Luminous transmittance vs. thickness for coatings of VO_xF_y and VO_2. The values of x and y are likely to vary somewhat among the samples owing to differences in preparation conditions. Curves were drawn for convenience. (From Ref. 144)

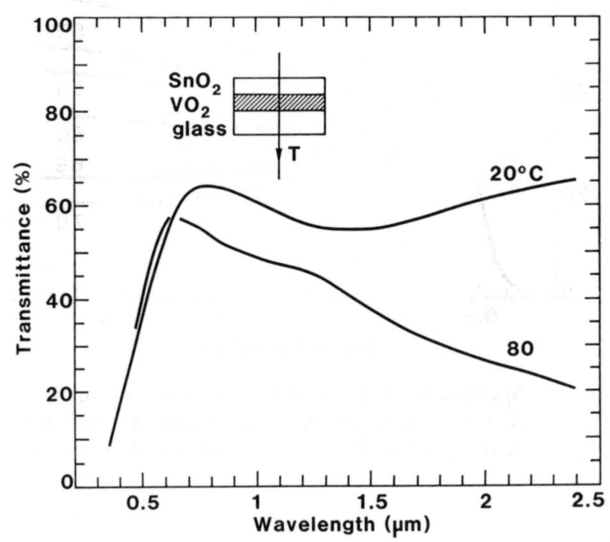

Fig. 28. Spectral normal transmittance for a research-type thermochromic smart window with a vanadium oxide/tin oxide tandem coating at two temperatures. (From Ref. 142).

G. Coatings with Angularly Dependent Transmittance

This section contains some examples of computed and measured transmittance through uniform and non-uniform metal-based coatings. For the case of uniform coatings, the transmittance is non-selective, i.e. $T(\theta) = T(-\theta)$, but the angular dependence can be strong for suitably designed multilayer stacks. Particularly interesting data are obtained by invoking more than one metal layer,[35] in which case the coating design becomes somewhat akin to that of a Fabry-Perot etalon.[61] Figure 29 shows two examples of T_{sol} (θ) for five-layer coatings consisting of two 12-nm-thick silver layers and three SiO_2 layers. Tabulated optical constants[62] pertinent to the bulk materials were used in the computations. When the SiO_2 thickness is 120 nm, there is a monotonic decrease of T_{sol} with increasing θ. For 170-nm-thick SiO_2 layers, the angular dependence is more interesting and goes from ~ 23 % at normal incidence to as much as ~ 58 % at 60° angle of incidence. Further work is needed to explore the limits of the angularly dependent optical properties in multilayer window coatings with more than one metal layer.

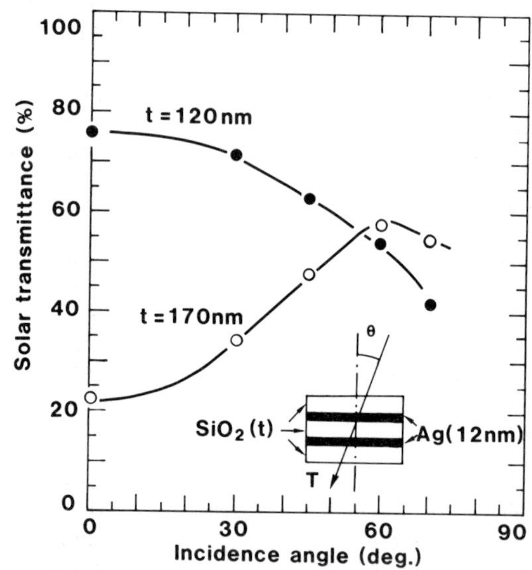

Fig. 29. Computed angularly dependent solar transmittance for $SiO_2/Ag/SiO_2/Ag/SiO_2$ coatings with two values of the SiO_2 thickness (t). The coating configuration is shown in the inset. Open and filled circles show computed data. Curves were drawn only for convenience. (From Ref. 35).

Angular selectivity, can occur if a coating - or more generally at least one of the layers in a multilayer stack - is non-homogeneous and has an optical axis which deviates from the normal to the coating. In order to be more specific, and point out some general results for angular selectivity, we consider a collimated light beam incident onto a flat uniform substrate with a coating of a material characterized by a unique off-normal optical axis. The situation is illustrated schematically in Fig. 30. The coating is taken to consist of identical cylindrical columns. A microstructure approaching this model can be realized in samples produced by special vacuum deposition techniques, as will be discussed shortly.

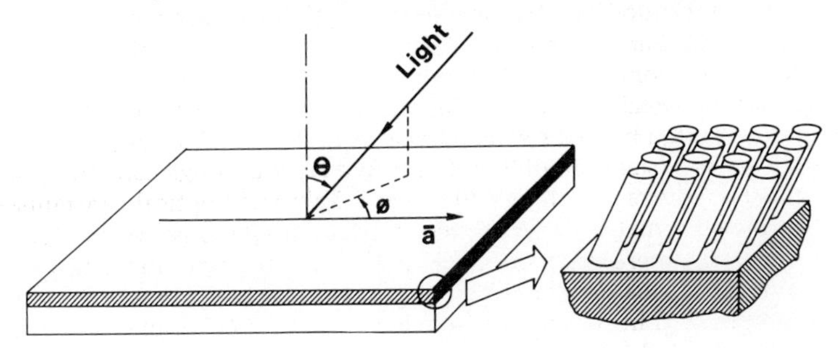

Fig. 30. The left-hand part defines the geometry for a light beam incident onto a coating of a uniaxial material. The right-hand part shows a schematic model for a columnar microstructure. (From Refs. 35 and 145).

The optical properties are conveniently described with regard to a vector a in the surface plane. One can then specify the incident light beam by its polar angle θ and its azimuthal angle ϕ. The vector a is chosen so that $T(\theta, \phi = 90°) = T(\theta, \phi = 270°)$; other orientations of the light beam yield $T(\theta, \phi) \neq T(\theta, \phi + 180°)$. In general, the difference between the transmittance values in the inequality is largest at $\phi = 0$, i.e., for light incident in the plane spanned by a and the surface normal. This particular configuration leads to a simple criterion for angular selectivity, which can be written for s polarization (electric field vector perpendicular to the incidence plane) and p polarization (electric field vector in the incidence plane) as $T_s(\theta) = T_s(-\theta)$ and $T_p(\theta) \neq T_p(-\theta)$. Here the sign convention $+\theta$ $(-\theta)$ denotes light having a propagation vector with a component opposite (parallel) to a. It is evident that angular selectivity is most pronounced for light with predominant p polarization.

Experimentally it is possible to achieve a microstructure resembling the one in the right-hand part of Fig. 30 by special etching techniques and, more interestingly, by vacuum deposition with the impinging beam having an angle α that deviates from the substrate normal. This experimental configuration is referred to as oblique angle deposition and is possible both with evaporation[35,37,38,145,146] and sputtering.[147] The relation between deposition angle and column orientation is often given by a "tangent rule", viz.[148]

$$\tan\beta = (1/2)\tan\alpha\,, \tag{6}$$

where β is the angle between the substrate normal and the symmetry direction for the columns (i.e., the column tilt). The general validity of Eq. (6) is questionable, though.[149,150]

Studies of angularly selective transmittance through obliquely evaporated chromium coatings have been conducted recently.[35,145] Figure 31 shows that both T_{lum} (θ) and T_{sol} (θ) increase monotonically with increasing angle until θ ≈ + 60° is reached. Beyond this angle the integrated transmittance drops. Specifically, T_{lum} goes from ~ 21 % at θ = - 60° to ~32 % at θ = + 60°, and T_{sol} goes from ~27 % at θ = – 60° to ~ 37 % at θ= + 60°. These results show clearly that angularly selective transmittance, of interest for energy-efficiency, can indeed be obtained. Data for aluminium and tantalum coatings are given in Ref. 38.

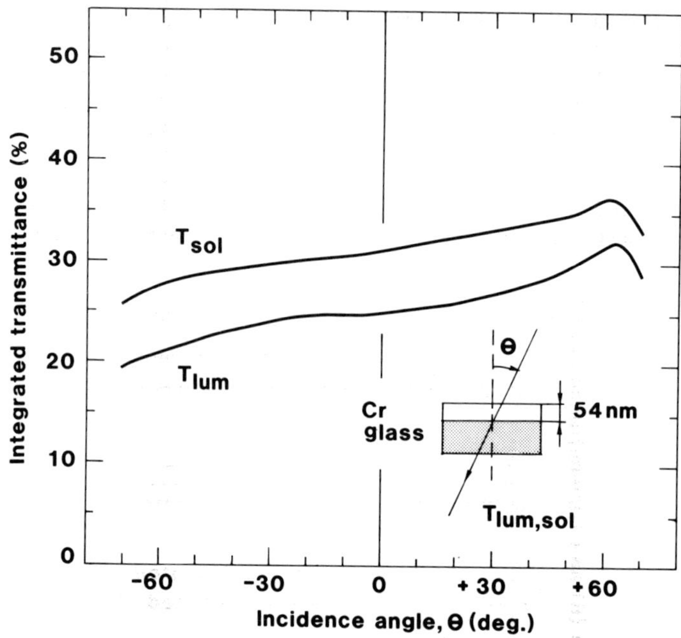

Fig. 31. Angularly dependent luminous and solar transmittance determined as sketched in the inset. (From Refs. 35 and 145).

H. Antireflection Treatments

It was pointed our earlier that each surface of an ordinary window glass produces ~ 4 % reflection, and that coated glass surfaces, in general, have a reflection that is even higher. This reflection is often undesirable both in terms of energy efficiency and since it can cause visually disturbing effects. A solution to this problem is to apply an antireflection (AR) treatment, which can invoke a layer of refractive index ~ 1.4 and thickness ~ 0.1 μm on top of the (coated) glass. Other solutions can be obtained by employing a chemical etching of the glass surface or by having thick polymeric coatings.

Dielectric AR layers of MgF_2 etc. have been produced by evaporation and used on lenses and other optical components for many years.[31] These layers are hard and chemically inert but do not lend themselves to efficient large-scale sputter deposition. In fact, until recently there were no durable low-refractive-index coatings which could be applied by high-rate sputtering. Currently the situation is changing, and a range of newly developed metal oxyfluoride coatings have shown very promising results.[44-46,151]

Aluminium oxyfluoride layers have been produced by high-rate sputtering onto glass with and without surface coatings.[44,45] This type of AR layer is stable in a humid atmosphere, under ultraviolet irradition, and during heating to high temperatures in air. Figure 32 shows spectral transmittance and reflectance for a 1 mm thick glass with and without an AR layer on one side. It is seen that T_{lum} is increased by a few percent and that R_{lum} is strongly decreased by the aluminium oxyfluoride. Figure 33 reports on another application in which a 0.08 μm thick aluminium oxyfluoride AR layer is put on 4 mm floatglass having a commercial $Bi_2O_3/Ag/Bi_2O_3$ coating. The full multilayer structure is characterized by $T_{lum} \approx 84$ %, $T_{sol} \approx 67$ % and $R_{lum} \approx 4$ %. Before the AR layer was applied, the coated glass had $T_{lum} \approx 77$ %, $T_{sol} \approx 61$ % and $R_{lum} \approx 11$ %. Hence the AR treatment enhances T_{lum} by 7 % and T_{sol} by 6 %, and decreases R_{lum} by 7 %.

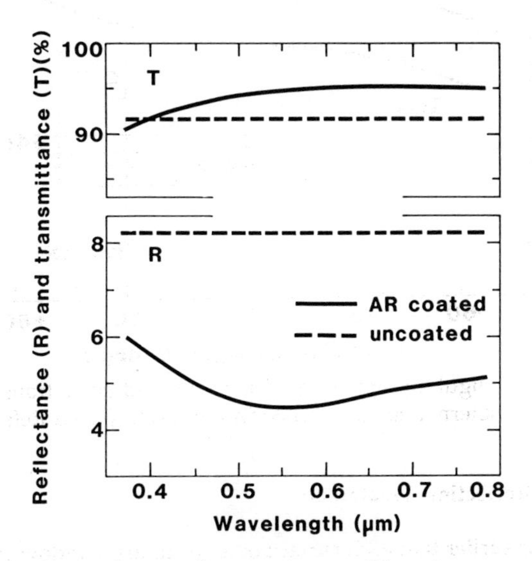

Fig. 32. Spectral normal transmittance (T) and near normal reflectance (R) measured for uncoated (dashed curves) and aluminium oxyfluoride coated (solid curves) glass. (From Ref. 44).

Aluminium oxyfluoride can be used also for AR treatment of doped oxide semiconductor coatings. However, for sputter-deposited coatings of those materials it may be more convenient to antireflect by an oxyfluoride based on the same material as the one in the oxide. A recently studied example is an In_2O_3:Sn

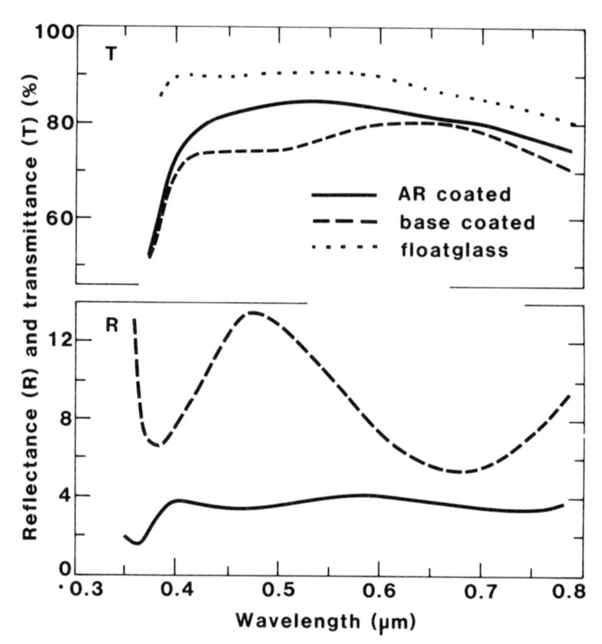

Fig. 33. Spectral normal transmittance (T) and near normal reflectance (R) measured for uncoated floatglass (dotted curve), for the same glass with a $Bi_2O_3/Ag/Bi_2O_3$ coating (dashed curves) and after application of a final aluminium oxyfluoride layer (solid curves). The base coated glass corresponds to the product I plus Neutral, supplied by Interpane, Germany. (From Ref. 44).

coating with an antireflecting indium-tin oxyfluoride overlayer.[151] The AR treatment increased T_{lum} from 82 to 89 % and decreased R_{lum} from 13 to 5 %.

Dielectric AR layers tend to enhance E_{therm} for noble-metal based as well as for doped oxide semiconductor coatings, but the effect is on the ~ 1 % level, which is of little practical significance. Further, the AR layers can be used to avoid iridescence[81,152] - particularly for specified viewing conditions - and to make light scattering (haze) less apparent.[153]

Another possibility to antireflect glass is by etching its surfaces with fluosilicic acid so that a porous silica layer, with low refractive index and well-defined thickness, is formed.[154-157] Figure 34 shows reflectance of 3-mm-thick floatglass with and without a double-sided AR treatment. It is seen that the reflectance can be very low over a large part of the luminous spectral range. The silica layer is almost non-absorbing, and hence there is a concomitant increase in T_{lum}. The AR layers can be created by liquid-phase and vapour-phase etching; potentially both of these are low-cost techniques suitable for large-area windows.

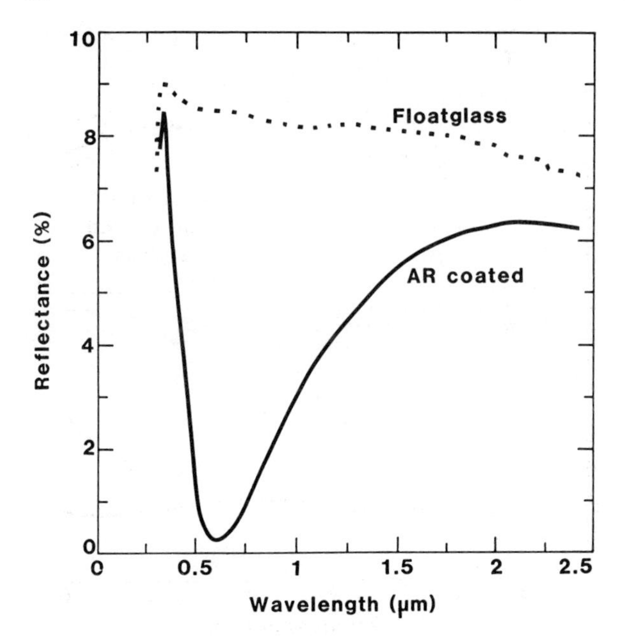

Fig. 34. Spectral near-normal reflectance for floatglass (dotted curve), and for such glass etched in fluosilicic acid (solid curve). (From Ref. 156).

V. MATERIALS INTERPOSED BETWEEN THE WINDOW PANES

A. Gases

Hermetically sealed doubly, or multiply, glazed windows make it possible to incorporate a gas other than air between the glass panes. In this way one can alter the heat transfer by thermal radiation, conduction and convection and thereby enhance the energy efficiency. The possible improvements are illustrated in Figure 35, which gives k-values measured as a function of the distance between two glasses for air and five different chemically well specified gases.[158] The data were recorded at the center of a ~ 0.6 m^2 vertical square test window with gas spacings of 6, 9, 12, 16 and 20 mm. The upper set of curves refers to normal uncoated glass with $E_{therm} \approx 85$ %. For air, argon (Ar), and carbon dioxide (CO_2), there is a monotonic drop in the k-value with increasing spacing s, whereas for sulfur hexafluoride (SF_6), Freon 12 (Cl_2CF_2) and sulfur dioxide (SO_2), the k-value is much less dependent on s. At s = 12 mm, which is a common value for sealed windows, air, argon, and CO_2 yield k-values of 3.0, 2.8 and 2.6 Wm^{-2} K^{-1}, respectively. Hence the improvement given by the gas is on the ~ 10 % level for uncoated glass.

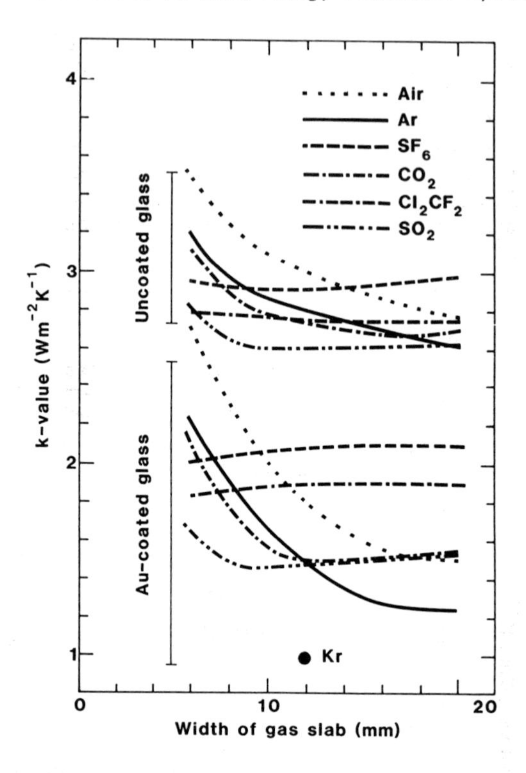

Fig. 35. k-value vs. glass spacing measured for double glazed test windows. Variously dashed and dotted curves refer to the shown gases used as fillings between the glass panes. The upper set of curves was recorded with uncoated glass. The lower set of curves corresponds to one of the glass surfaces having a gold coating. A single data point is shown for krypton gas. (From Ref. 158).

Gas fillings are conveniently combined with low-emittance coatings based on noble metals or doped oxide semiconductors.[158,159] The lower part of Fig. 35 shows results obtained with a window for which one of the surfaces facing the gas enclosure was gold coated and had $E_{therm} \approx 6.5$ %. When the radiative component to the heat transfer is minimized, the choice of a gas with low thermal conduction and convection becomes critical, and the relative improvement possible by selecting a proper gas becomes enlarged. It is found that the k-value drops with increasing s for air and argon, and that the other gases give a more complicated behaviour or even a rise of the k-value with increasing s. At s = 12 mm one finds a k-value of 1.8 $Wm^{-2} K^{-1}$ for air - in good agreement with Fig. 2 - and 1.5 $Wm^{-2} K^{-1}$ for argon. Thus an improvement on the 20 % level can be obtained by gas filling together with low emittance coated glass. For krypton (Kr) gas the k-value can be as low as 0.98 $Wm^{-2} K^{-1}$ at s = 12 mm. Xenon (Xe) is expected to be even better.

Gas mixtures are also of interest for obtaining low k-values.[158] Figure 36 shows that a combination of argon and SF_6 can give 2.7 $Wm^{-2}K^{-1}$ at ~ 30 vol. % SF_6 in the case of uncoated glass and s = 12 mm. For gold-coated glass, the k-value is almost unaffected by SF_6 mixtures up to ~ 20 vol. %. A mixing of SF_6 and air can also be of advantage. The addition of SF_6 leads to better sound insulation, which is another benefit.[160]

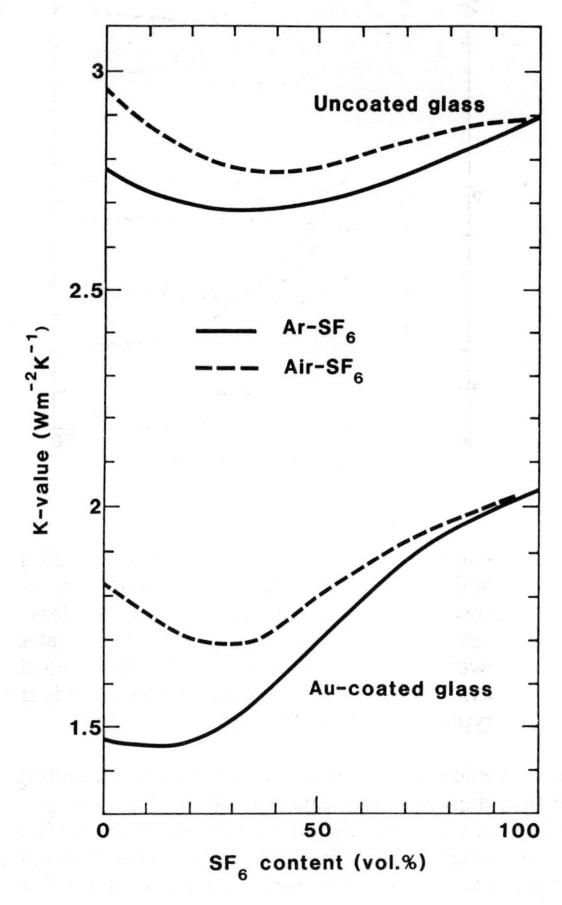

Fig. 36. k-value vs. relative SF_6 content measured for Ar-SF_6 gas mixtures. (From Ref. 158).

Recently attempts have been made to decrease the radiative heat transfer in windows by infrared-absorbing gases.[161] Many gases - including the earlier mentioned SF_6 and Cl_2CF_2 - have strong but narrow absorption bands in the $3 < \lambda < 50$ μm range, and by combining several component gases one can minimize the heat transfer. A lowest possible T_{therm} (0) of ~ 70 % was obtained across a 10 mm gas layer. Figure 37 shows the infrared transmittance spectrum for a mixture of four different freons and SF_6. Analogous results can be obtained by non-chlorinated gases, which are not harmful to atmospheric ozone.

The cost of gases is rougly 0.01 for argon, 0.1 for sulfur hexafluoride, 1 for krypton and 10 for xenon in units of USD per litre and assuming an annual consumption of < 1000 m^3 at the filling site. For krypton and xenon, global availability precludes large-scale uses. The cost of infrared-absorbing gas is ~ 0.2 USD per litre. The filling cost for a standard window is ~ 0.25 USD on the production line and ~1 USD off the production line.

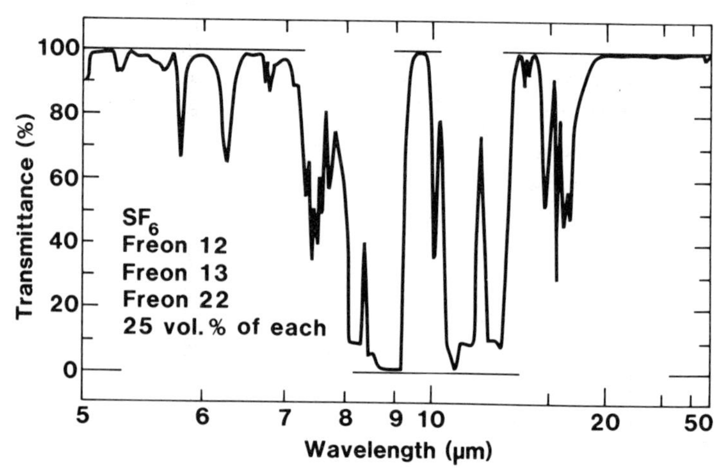

Fig. 37. Spectral normal transmittance through a 10-mm-thick layer of a gas mixture with the shown composition. (From Ref. 161).

B. Solid Transparent Insulation Materials

There is a wide range of solid materials which can be interposed between glass panes in order to improve the thermal insulation of windows. The earlier chapter on Transparent Insulation Materials covers this subject in detail, and the present brief discussion is included only for completeness. Figure 38 gives a convenient subdivision into four groups of (i) thin flexible polymer foils, (ii) polymer honeycomb materials, (iii) bubbles, foams and fibres, and (iv) inorganic microporous materials, especially silica aerogels. If the honeycomb cross-section is small compared to the cell length, one may speak of a capillary structure. Materials (i) and (iv) can be almost invisible to the eye, whereas materials (ii) and

(iii) cause strong reflecting effects. Hence the "macroporous" honeycombs, bubbles, foams and fibres do not give a good visual indoor-outdoor contact, but they may be used to provide lighting; their main application is thought to be in innovative solar collectors and wall claddings.[162] Materials (ii) and (iv) allow, in principle, very high solar energy throughput. Detailed reports on solid transparent insulation materials have been given by Pflüger[163] and Platzer,[164] as well as in an earlier chapter of this book.

Flexible foils can be mounted between glass panes by use of frames which maintain the material in permanent tension. Foils have their main application as a carrier for a coating devised for giving a high T_{lum}/T_{sol} ratio, as required for a warm climate, or for giving a high T_{sol} together with a low E_{therm}, as required for a cold climate. Thus one can create a light-weight multiple glazing in conjunction with energy efficient coatings by incorporating only two panes of uncoated glass. Clear polyester foil with low surface roughness is well suited as a substrate material for noble-metal based and doped oxide semiconductor coatings. Sputter technology is often applied in such a way that an entire roll of plastic web is placed in the vacuum tank and is wound over to another roll so that the material passes uniformly near the target(s). The technique is referred to as "roll coating". Normal plastics must remain at $< 100°$ C during the coating process.

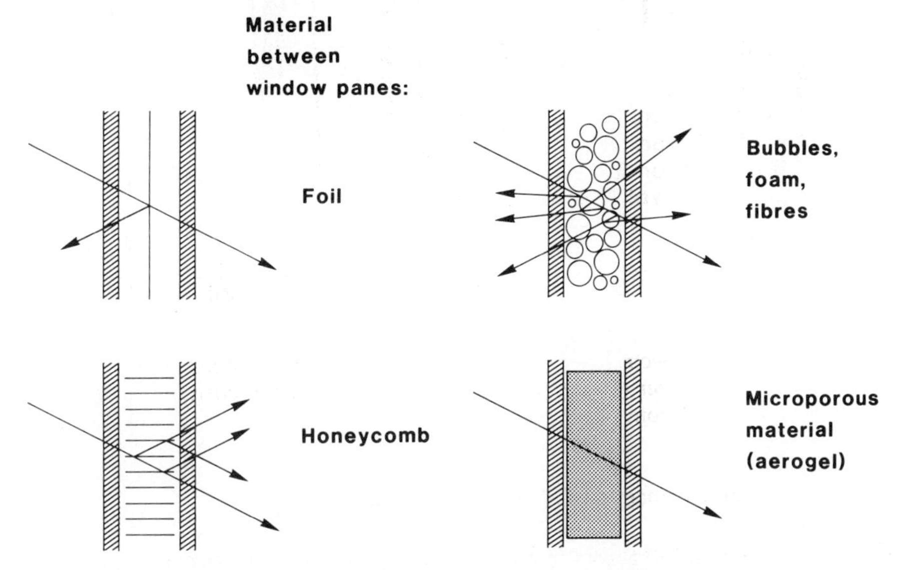

Fig. 38. Principles of four different types of solid transparent insulation materials placed between two glass panes. Arrows signify light rays. Reflections at the glass surfaces are not shown. The classification follows Platzer and Wittwer (Ref. 165).

Figure 39 shows transmittance and reflectance for polyester foil with three types of coatings. The upper part refers to a $SnO_2/Ag/SnO_2$ coating,[166] whose properties are similar to those for $TiO_2/Ag/TiO_2$ reported in Figure 12. Silver-based coatings with two chemically prepared dielectric layers have also been developed.[167] The middle and lower parts of Fig. 39 show data for doped In_2O_3 from Howson et al.[168] and doped ZnO from Jin and Granqvist,[169] respectively. Generally speaking, the properties are similar, though not quite as good, as for analogous coatings on glass (cf. Figure 17). Low-temperature techniques are available also for doped SnO_2.[170,171] The cost for coating plastic foil is comparable to the one for coating glass panes, at least for noble-metal based layers.

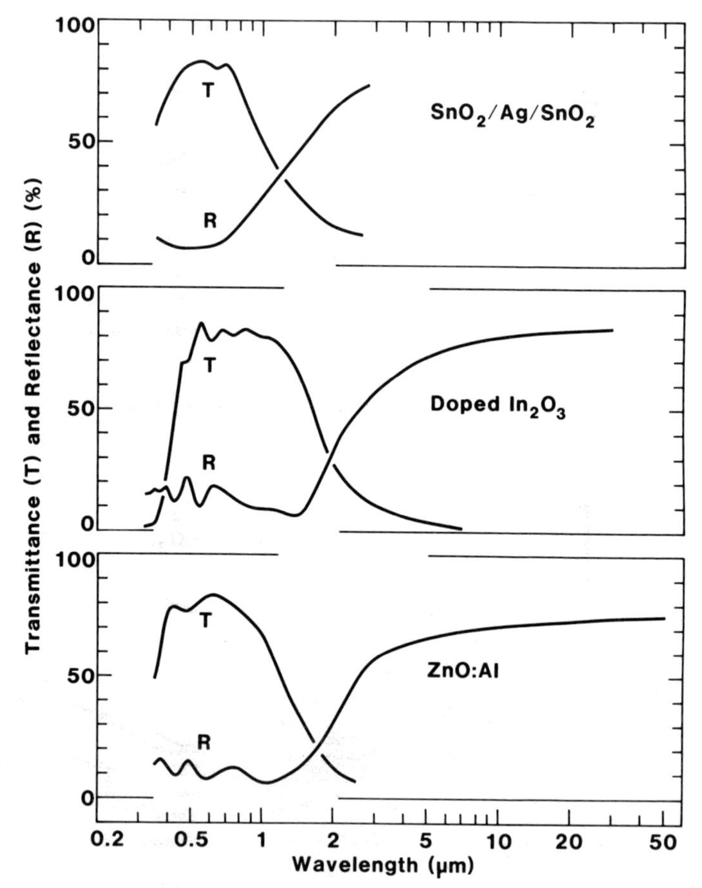

Fig. 39. Spectral normal transmittance and near-normal reflectance measured for clear polyester foil coated with $SnO_2/Ag/SnO_2$, doped In_2O_3, and ZnO:Al. Compiled from Refs. 166, 168 and 169.

We now turn to the class of macroporous materials, including honeycombs, capillaries, foams, bubbles, fibres, etc. Transmittance for diffuse solar irradiance, denoted T_{sol}^{d}, and k-value have been reported for numerous configurations in Refs. 172 and 173. Honeycombs of different types - made of polystyrol, polyamide, polyvinylchloride and polycarbonate - typically gave T_{sol}^{d} between 71 and 45 % and k-values between 1.3 and 0.9 $Wm^{-2}K^{-1}$ when the thickness was ~ 10 cm.

Figure 40 reports spectral transmittance at diffuse irradiation for the polycarbonate honeycomb.[162] The transmittance is large at $0.4 < \lambda < 1.6$ μm, which yields $T_{sol}^{d} \approx 71$ %. The thickness dependence of T_{sol}^{d} is of obvious interest for windows. Figure 41 shows representative results for polycarbonate capillaries with 1.7 mm diameter and for polymethylmethacrylate foam.[173] For the capillary materials, T_{sol}^{d} is seen to drop from ~ 73 % at small thickness to ~60 % at 10 cm thickness; the k-values were 1.45 and 0.79 $Wm^{-2}K^{-1}$ at 6 and 10 cm thickness, respectively.[172] The foam layers display a much stronger thickness dependence of T_{sol}^{d}, which drops below 50 % when the thickness exceeds ~ 2.5 cm. At 1.6 cm thickness, different foam types showed T_{sol}^{d} between 60 and 48 % and k-values between 3.1 and 3.6 $Wm^{-2}K^{-1}$.[172]

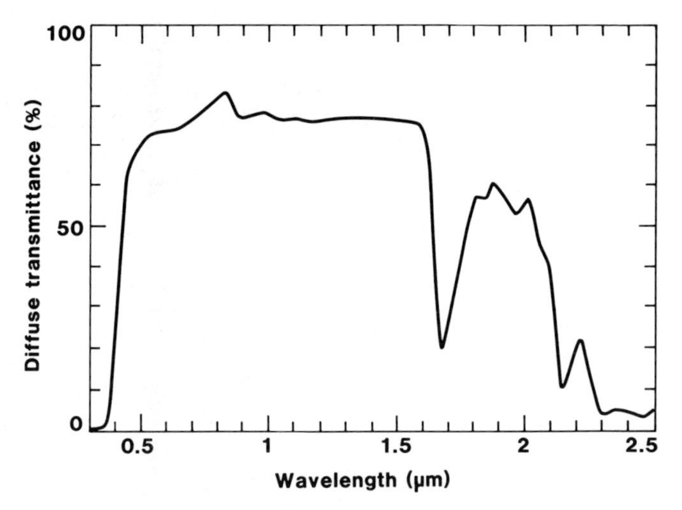

Fig. 40. Spectral transmittance for diffuse irradiance measured for a 10-mm-thick rectangular polycarbonate honeycomb constructed from foil of 60 μm average thickness. (From Refs. 162 and 174).

By going from a macroporous to a microporous structure - with inhomogeneities much less than the wavelengths of visible light - one can achieve a transparent solid material with superior thermal insulation.[175] A particularly interesting material can be obtained by supercritical drying of colloidal silica gel.[176,177] The ensuing substance, called a silica aerogel, consists of silica particles of size ~ 1 nm interconnected so that a loosely packed structure with pore sizes of 1 to 100 nm is formed. The aerogel density is 70 to 270 kg m^{-3}, compared with 2200 kg m^{-3} for

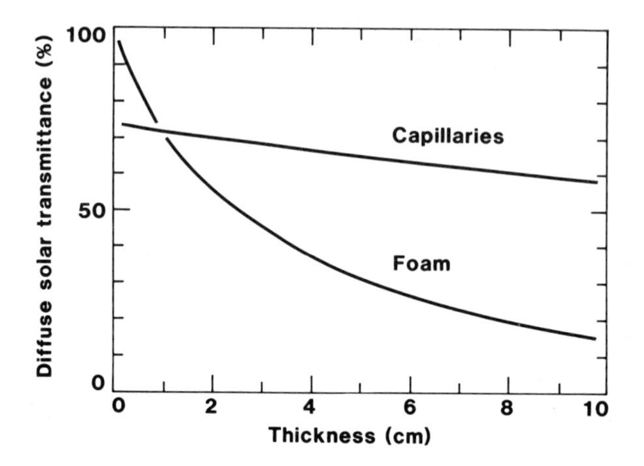

Fig. 41. Transmittance for diffuse solar irradiation vs. thickness measured for materials comprising polycarbonate capillaries and polymethylmethacrylate foam. (From Ref. 173).

non-porous silica glass, implying a porosity up to 97 %. Many properties of silica aerogels are compiled by Fricke.[178] Large-scale production is discussed in Refs. 179 and 180. Silica aerogel can be prepared both as transparent tiles and as a translucent granular material.

A high T_{lum} and T_{sol} together with a low k-value makes silica aerogel of great interest for energy-efficient windows with and without low-emittance coating.[181-185] Further, the mechanical rigidity permits its use as a spacer in evacuated windows. Figure 42 illustrates spectral transmittance for a 4 mm thick tile.[183] The transmittance exceeds 80 % at $0.6 < \lambda < 2.2$ μm but drops at $\lambda < 0.6$ due to scattering from density variations. Transmittance spectra for granular aerogels have been reported in Ref. 182. Figure 43 shows the thickness dependence of the transmittance, at direct and diffuse solar irradiation, through windows filled with granular silica aerogel.[174,182] As expected, the transmittance falls off with increasing thickness; at 1 cm one finds $T_{sol} \approx 69$ % and $T_{sol}{}^{d} \approx 55$ %. The k-value of this kind of material is shown in Fig. 44 as a function of the gas pressure in the aerogel and for two magnitudes of the emittance of the surrounding surfaces.[184] The measurements were done at ~ 40°C. With nitrogen gas at atmospheric pressure, the k-value is ~ 1.8 $Wm^{-2}K^{-1}$ for a 1.5 cm thick layer of granules. Decreasing the gas pressure makes the k-value drop. For pressures below 1 mbar the radiative heat exchange becomes of significance - particularly at $3 < \lambda < 5$ μm - and hence an enclosure with low emittance yields a depressed k-value. At 10^{-3} mbar, the k-value is ~ 0.9 for an emittance of 0.9 (representative of uncoated glass) and ~ 0.6 for an emittance of 0.05 (representative of low-emittance-coated glass). For 0°C one can extrapolate from data by Fricke et al.[184] that the corresponding k-values would be 0.7 and 0.5, respectively. Granular silica aerogel produced in large quantities is expected to cost approximately 1.8 USD per litre.

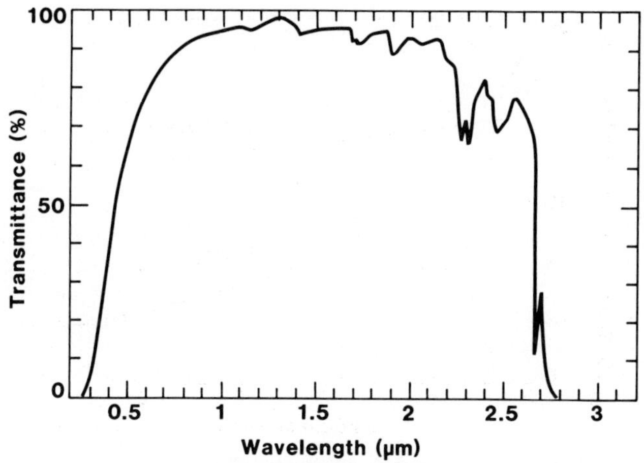

Fig. 42. Spectral transmittance measured for a 4 mm thick silica aerogel tile. The material was heat treated at 500°C in order to enhance the transmittance by the removal of organic residues and water. (From Ref. 183).

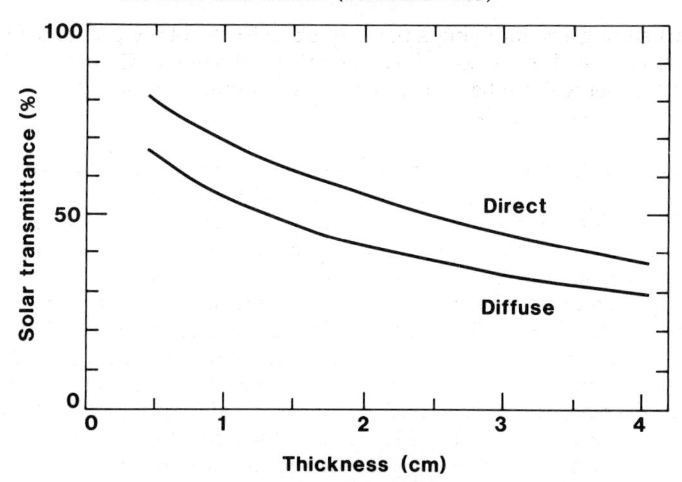

Fig. 43. Solar transmittance vs. thickness measured for granular silica aerogel samples. The curves refer to direct incidence (i.e., T_{sol}) and diffuse incidence (i.e., T_{sol}^{d}). (From Refs. 174 and 182).

Instead of having aerogel spacers in evacuated windows, it is possible to use small spherical glass supports. Such a window, with low emittance coating and evacuated to $< 10^{-3}$ mbar, can yield a k-value of 0.6 $Wm^{-2}K^{-1}$ in a compact (< 1 cm thick) and light-weight (~ 14 kg m^{-2}) design.[186,187] Laser edge sealing of the evacuated window is a promising technique.[188]

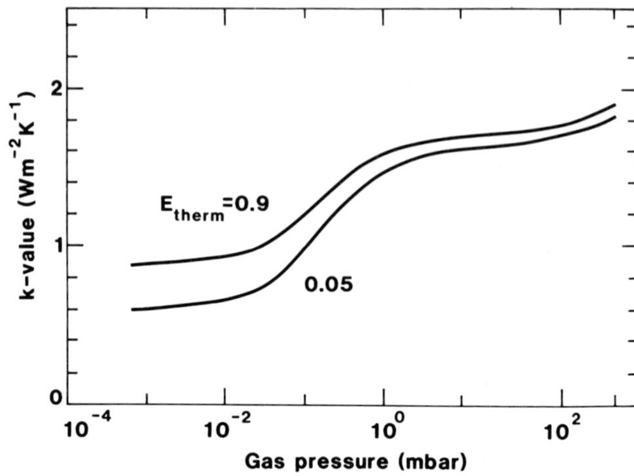

Fig. 44. k-value vs. pressure of nitrogen gas for a 1.5 cm thick layer of silica aerogel. The curves refer to the shown magnitudes of the thermal emittance for the surfaces surrounding the areogel; they were drawn through individual data points and may be somewhat uncertain as regards details. The aerogel is characterized by a density of 230 kg m^{-3} and a granule size of 3 mm. (From Ref. 184).

C. Chromogenic Materials

Chromogenic coatings on glass were discussed above with a focus on electrochromics-based and thermochromic smart windows. This section discusses some recent work on "thick" materials which allow dynamic throughput of radiant energy when interposed between window panes. Specifically, covered are photochromic, thermochromic, and electrically controlled liquid-crystal-based materials.[18]

Plastic sheet or foil with *photochromic* additives offer interesting possibilites,[189,190] and detailed results have been given for spirooxazine in a host of cellulose acetate butyrate. The material can be produced by injection molding. Figure 45 shows darkening and clearing dynamics of T_{lum} for a 1.5 mm thick sheet at 20° C. It is seen that solar exposure makes T_{lum} drop from 82 to 23 % in about one minute. Clearing is slower, and it takes about 20 minutes to regain full transmittance. The dynamics are much more rapid than for photochromic glass (cf. Fig. 5). Spectral transmittance in the $0.35 < \lambda < 0.75$ µm range is illustrated in Fig. 46 for the sheet in fully cleared and in partially darkened states. Darkening is associated with absorption at $0.5 < \lambda < 0.7$ µm, which is analogous to the absorption in photochromic glass (cf. Fig. 6). The spirooxazine-induced photochromism is temperature dependent to an undesirable extent, and at 30 and 40° C the fully darkened state for the material of Fig. 45 corresponds to $T_{lum} \approx 40$ and 70 %, respectively.

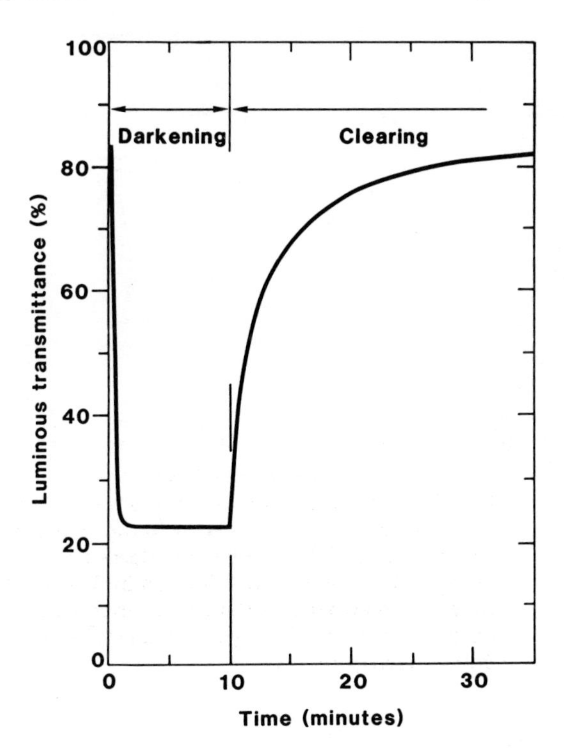

Fig. 45. Luminous transmittance vs. time during darkening and clearing of a plastic sheet with photochromic additives. (From Ref. 189).

Thermochromic control of radiation throughput can be accomplished by different kinds of materials. Polymeric "cloud gels" are well known in this context.[191-193] Clouding - i.e., transition to a diffusely scattering state - can set in above a certain temperature due to a reversible thermochemical dissolution and a thermally induced modification in the length of the polymer molecules. The cloud point can be regulated to within 1.5°C in the 9 to 90°C range.[193] Figure 47 shows transmission of direct plus scattered radiation through a 1 mm thick cloud gel layer interposed between two glass panes. Both the luminous and the solar transmittance drop by ~ 50 % when the cloud point is exceeded. The material has to be used in a sealed window. Its cost in a window can be less than 1.80 USD/m^2. On the negative side, we note that if thermal gradients exist over a cloud-gel-containing window these may manifest themselves as areas with clear and cloudy appearances. Apart from cloud gels, thermochromic liquid/fibre composites have been studied recently.[194]

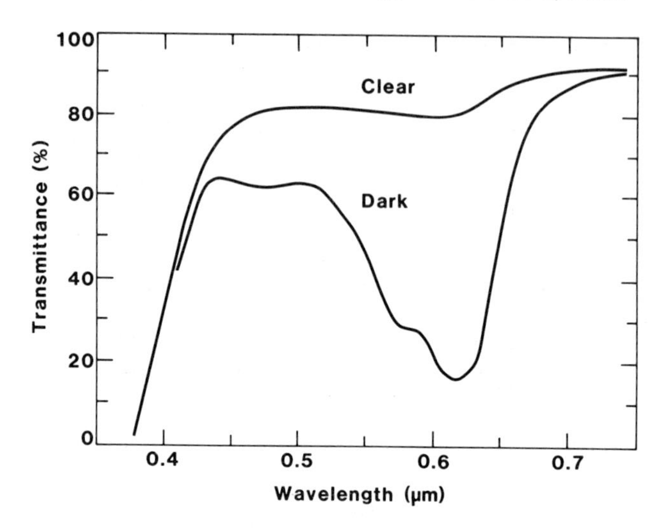

Fig. 46. Spectral normal transmittance for a plastic sheet with photochromic additives in clear and dark states. (From Ref. 189).

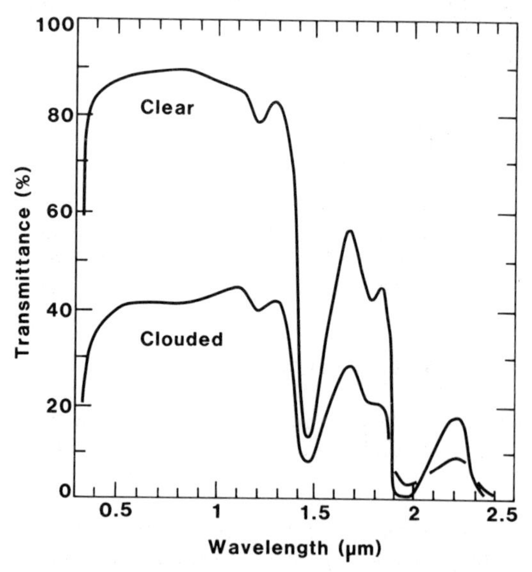

Fig. 47. Spectral total (direct and scattered) transmittance through a cloud gel layer between glass panes. The curves refer to a low-temperature clear state and a high-temperature clouded state. (From Ref. 193).

Electrically controlled *liquid-crystal-based materials* can switch between a transparent state and a highly scattering state in a way principally similar to the one for cloud gels.[195,196] A material which recently has been commercialized consists of a polymer layer with micrometre-sized cavities containing a nematic liquid crystal. This layer is laminated between two In_2O_3:Sn-coated polyester foils serving as extended electrodes. Applying a voltage over the electrodes, one can align the liquid crystal molecules and thereby obtain refractive-index-matching between the inclusions and the surrounding polymer matrix. This represents the transparent state of the material. In the absence of an electric field, the liquid crystal molecules become randomly oriented, which creates refractive-index-mismatch and hence strong scattering. Figure 48 shows transmission of direct and scattered radiation through this type of material in the unpowered state and when a 60 Hz 100 V square wave is applied. A switching between 82 and 62 % transmission is possible. An electric power of < 20 Wm^{-2} is needed for operating the window.

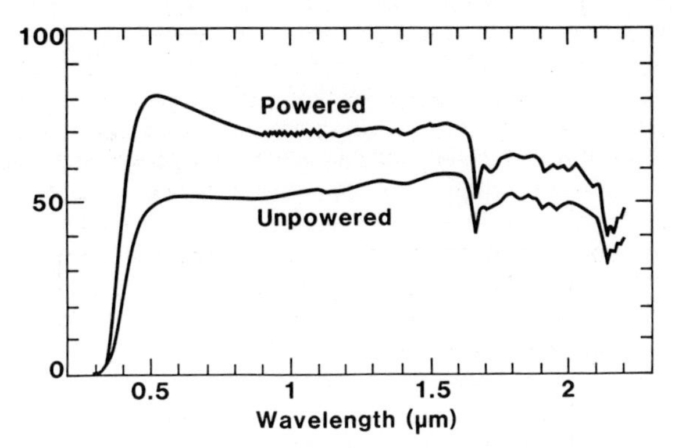

Fig. 48. Spectral total (direct and scattered) transmittance through a liquid-crystal-based material laminated onto 3.2-mm-thick window glass. The curves refer to unpowered (opaque) and powered (transparent) states. (From Ref. 196).

VI. CONCLUSIONS AND OUTLOOK

This chapter has embraced a large number of possibilities for achieving the energy efficiency of architectural windows. This multiplicity of options is easily understood since

(i) energy efficiency means different things in different climates,
(ii) there are several principle approaches to energy efficiency which can be utilized either separately or in conjuction, and
(iii) many different materials may be useful for each climate and approach to energy efficiency.

An energy efficient window must include at least two *glass* panes. Floatglass is a rather standardized product but it should be kept in mind that special qualities with low Fe_2O_3 content can yield a significant improvement of the solar energy transmission.

Thin *coatings* can modify the radiative properties of the glass surfaces in many different ways. Thus one can use noble-metal based coatings for blocking the transmission of infrared solar radiation thereby cutting down the heating, or one can use noble-metal based or doped oxide semiconductor coatings for combining high transmission of solar radiation with low emission of thermal radiation. These coatings are produced commercially on a large scale. Noble-metal based coatings must be used in hermetically sealed environments, whereas doped oxide semiconductors are inert and rugged. Iridescence and haze may be problems for doped oxide semiconductor coatings, albeit not unsurmountable ones. Coatings with dynamic properties - to be used in smart windows - are subject to vigorous research efforts and are rapidly approaching a stage where commercialization is feasible. All-solid-state electrochromics-based coatings are of particular interest. Laboratory studies indicate that their transmittance of visible light and solar energy can be altered gradually and reversibly between ~ 10 and ~80 % by electric pulses; the energy requirement is very small. Thermochromic coatings is another, more remote, possibility. Novel antireflection coatings seem to be able to increase the transmittance and decrease the reflectance of window glass with and without various types of coatings.

Materials interposed between the window panes can diminish the heat transfer due to conduction and convection. Gases, particularly argon and SF_6, as well as low-emittance-coated polymer foils are well understood in this context. Low-density silica aerogel tile is a novel material capable of giving an extremely low heat transfer.

The k-value is one of the parameters which govern energy efficiency. It is illustrative to follow its decrease through the following series of measurements. Starting with a singly glazed window aperture one has a k-value of ~ 6 $Wm^{-2}K^{-1}$; introducing double glazing leads to ~ 3 $Wm^{-2}K^{-1}$; adding a low emittance coating gives ~ 1.8 $Wm^{-2}K^{-1}$, and adding a suitable gas in the space between the panes yields ~ 1.5 $Wm^{-2}K^{-1}$. With a moderate vacuum between the panes, which requires a spacer of for example silica aerogel, one may reach an ultimate k-value of ~ 0.5 $Wm^{-2}K^{-1}$.

Current research and development on energy efficient windows is likely to lead to significant changes in fenestration. An assessment of these must be a subjective one - but will nevertheless be attempted. Noble-metal based coatings are presently produced with properties approaching the theoretical optimum. Research on technologies for making thinner continuous metal layers than those now available may lead to a marginal improvement in transmittance and drop in cost. Doped oxide semiconductor coatings, produced by large-area techniques, do not yet have optical properties which quite match the theoretical limits. Refinements in deposition technology, particularly in advanced spray pyrolysis, may lead to low-cost coatings with < 1 % luminous absorptance, no discernible haze, low thermal emittance, and electrical resistivity down to ~ 10^{-4} Ω cm. Novel ZnO-based coatings are of interest since they can block some of the ultraviolet solar radiation which may otherwise cause degradation of plastics and textiles. Electrochromics-based coatings will remain as subjects for intense research, and several new transition metal oxide coatings and solid electrolytes (particularly polymeric ones) are being investigated. This will give a broad basis for selecting an optimum device construction. Designs with two glasses, each having a two-layer coating, laminated together by a transparent polymeric electrolyte are easier to accomplish than five-layer designs. Recent results with laminated devices (cf. Figure 24) point towards future large scale electrochromics-based smart windows. However, a final assessment of their usefulness must await long-term environmental testing with a focus on irreversible photochromic effects. For thermochromic coatings there is a need for more basic research. Low-refractive-index antireflection coatings will be studied with a focus on the optimization of sputter conditions for various metal oxyfluorides.

Gas fillings which produce a low heat transfer are well known and widely used. Judging from their physical and chemical properties, krypton and xenon are the best, but global availability precludes large-scale uses. Silica aerogel is of considerable interest, but research into low-cost technologies for large-scale production of transparent tiles or plates is needed before its practicality can be ascertained.

The most far-reaching implication of the research is that the role of the window may change. Instead of being a passive building component, which commonly lets in or out too much energy, it can become an active part which regulates the inflow of radiant energy in response to dynamic needs while remaining thermally well insulated. Such a development is in keeping with today's tendencies towards "intelligent buildings" equipped with advanced measuring and control technologies.

Acknowledgement: Many people have read the manuscript of this paper in its various versons. In particular we acknowledge the helpful comments by Dr. G.A. Niklasson, Chalmers University of Technology and Univeristy of Gothenburg, Sweden, Dr. C.-G. Ribbing, University of Uppsala, Sweden, Dr. J.S.E.M. Svensson, CoAT AB, Sweden, and Dr. V. Wittwer, Fraunhofer Institut für Solare Energiesysteme, Germany.

REFERENCES

1. C.G. Granqvist, in *Physics of Non-Conventional Energy Sources and Material Science for Energy*, edited by G. Furlan, N.A. Mancini, A.A.M. Sayigh and B.O. Seraphin (World Scientific, Singapore, 1987), p. 217.
2. C.G. Granqvist, in *Physics and Technology of Solar Energy*, edited by H.P. Garg et.al. (Reidel, Dordrecht, The Netherlands, 1987), Vol. 2, p. 191.
3. C.G. Granqvist, in *Electricity: Efficient End-Use and New Generation Technologies, and Their Planning Implications*, edited by T.B. Johansson, B. Bodlund and R.H. Williams (Lund University Press, Lund, Sweden, 1989), p. 89.
4. C.G. Granqvist, *Spectrally Selective Surfaces for Heating and Cooling Applications* (SPIE Optical Engineering Press, Bellingham, USA, 1989).

5. C.G. Granqvist, in *Material Science and the Physics of Non-Conventional Energy Sources*, edited by G. Furlan, D. Nobili, A.A.M. Sayigh and B.O. Seraphin (World Scientific, Singapore, 1989), p. 1.
6. C.M. Lampert, in *Physics of Non-Conventional Energy Sources and Materials Science for Energy*, edited by G. Furlan, N.A. Mancini, A.A.M. Sayigh and B.O. Seraphin (World Scientific, Singapore, 1987), p. 143.
7. C.M. Lampert, in *Materials Science and the Physics of Non-Conventional Energy Sources*, edited by G. Furlan, D. Nobili, A.A.M. Sayigh and B.O. Seraphin (World Scientific, Singapore, 1989), p. 45.
8. K. Ya. Kondratyev, *Radiation in the Atmosphere* (Academic, New York, 1969).
9. M.P. Thekaekara, in *Solar Energy Engineering*, edited by A.A.M. Sayigh (Academic, New York, 1977), p. 37.
10. T.S. Eriksson and C.G. Granqvist, Appl. Opt. 21, 4381 (1982).
11. G. Wyszecki and W.S. Stiles, *Color Science*, second edition (Wiley, New York, 1982), p. 256.
12. K.M. Hartmann, in *Biophysik*, edited by W. Hoppe, W.L. Lohmann, H. Markl and H. Ziegler (Springer, Berlin, 1977), p. 214.
13. C.G. Granqvist, Appl. Opt. 20, 2606 (1981).
14. M. Rubin, R. Creswick and S. Selkowitz, in *Proc. Fifth National Passive Solar Conf.* (Pergamon, New York, 1980), p. 990.
15. B. Karlsson, T. Karlsson and C.-G. Ribbing, J. Thermal Insulation 7, 111 (1983).
16. E. Bollin and J. Schmid, in *Advances in Solar Energy Technology*, edited by W.H. Bloss and F. Pfisterer (Pergamon, Oxford, 1988), p. 3503.
17. B. Kunz and M. Kunz, in *Advances in Solar Energy Technology*, edited by W.H. Bloss and F. Pfisterer (Pergamon, Oxford, 1988), p. 3545.
18. C.M. Lampert and C.G. Granqvist, editors, *Large-Area Chromogenics: Materials and Devices for Transmittance Control* (SPIE Optical Engineering Press, Bellingham, 1990).
19. J.S.E.M. Svensson and C.G. Granqvist, Solar Energy Mater. 12, 391 (1985).
20. R. Siegel and J.R. Howell, *Thermal Radiative Heat Transfer*, second edition (McGraw-Hill, New York, 1981).
21. F. Kreith and W.Z. Black, *Basic Heat Transfer* (Harper & Row, New York, 1980).
22. W.M. Rohsenow, J.P. Hartnett and E.N. Ganic, *Handbook of Heat Transfer Fundamentals*, second edition (McGraw-Hill, New York, 1985).
23. J. Lohrengel, Glastechn. Ber. 43, 493 (1970).
24. R.J. Araujo, Contemp. Phys. 21, 77 (1980).
25. H.J. Hoffman, in *Photochromic Materials and Systems*, edited by H. Dürr and H. Bouas-Laurent (Elsevier, Amsterdam, 1990).
26. E.L. Swarts and J. Pressau, J. Am. Ceram. Soc. 48, 333 (1965).
27. G.S. Meiling, Phys. Chem. Glasses 14, 118 (1973).
28. G. Glimeroth and K.-H. Mader, Angew. Chem. Internat. Edit. 9, 434 (1970).
29. L.I. Maissel and R. Glang, editors, *Handbook of Thin Film Technology* (McGraw-Hill, New York, 1970).
30. J.L. Vossen and W. Kern, *Thin Film Processes* (Academic, New York, 1978).
31. H.K. Pulker, *Coatings on Glass* (Elsevier, Amsterdam, 1984).
32. L. Eckertova, *Physics of Thin Films*, second edition (Plenum, New York, 1986).

33. W.D. Westwood, Phys. Thin Films 14, 1 (1989).
34. C.G. Granqvist, in *Solar Optical Materials*, edited by M.G. Hutchins (Pergamon, Oxford, 1988), p. 59.
35. G. Mbise, G.B. Smith, G.A. Niklasson and C.G. Granqvist, Proc. Soc. Photo-Opt. Instrum. Engr. 1149, 179 (1989).
36. G.B. Smith, Opt. Commun. 71, 279 (1989).
37. G. Mbise, T. Otiti and R.T. Kivaisi, in *Energy and Environment into the 1990s*, edited by A.A.M. Sayigh (Pergamon, Oxford, 1990), Vol. 3, p. 1411.
38. G.B. Smith, R.J. Ditchburn and M.W. Ng, in *Energy and Environment into the 1990s*, edited by A.A.M. Sayigh (Pergamon, Oxford, 1990), Vol. 3, p. 1406.
39. C.M. Lampert, Solar Energy Mater. 11, 1 (1984).
40. S.M. Babulanam, T.S. Eriksson, G.A. Niklasson and C.G. Granqvist, Solar Energy Mater. 16, 347 (1987).
41. H. Marquez, J.M. Rincon and L.E. Celaya, Thin Solid Films 189, 139 (1990).
42. M.R. Jacobson, P.D. Hillman, A.L. Phillips and U.J. Gibson, Proc. Soc. Photo-Opt. Instrum. Engr. 428, 57 (1983).
43. C.F. Hickey and U.J. Gibson, J. Appl. Phys. 62, 3912 (1987).
44. G.L. Harding, Solar Energy Mater. 12, 169 (1985).
45. G.L. Harding, I. Hamberg and C.G. Granqvist, Solar Energy Mater. 12, 187 (1985).
46. G.L. Harding, Thin Solid Films 138, 279 (1986).
47. E. Valkonen, B. Karlsson and C.-G. Ribbing, Solar Energy 32, 211 (1984).
48. C.-G. Ribbing and E. Valkonen, Proc. Soc. Photo-Opt. Instrum. Engr. 652, 166 (1986).
49. E. Valkonen, C.-G. Ribbing and J.E. Sundgren, Proc. Soc. Photo-Opt. Instrum. Engr. 652, 235 (1986).
50. A.G. Spencer, M. Georgson, C.A. Bishop, E. Stenberg and R.P. Howson, Solar Energy Mater. 18, 87 (1988).
51. Y. Claeson, M. Georgson, A. Roos and C.-G. Ribbing, Solar Energy Mater. 20, 455 (1990).
52. J.C.C. Fan and F.J. Bachner, Appl. Opt. 15, 1012 (1976).
53. E. Valkonen and B. Karlsson, Energy Res. 11, 397 (1987).
54. E. Kusano, J. Kawaguchi and K. Enjouji, J. Vac. Sci. Technol. A 4, 2907 (1986).
55. S.J. Nadel, J. Vac. Sci. Technol. A 5, 2709 (1987).
56. M. Georgson, A. Roos and C.-G. Ribbing, to be published.
57. J.L. Vossen, Phys. Thin Films 9, 1 (1977).
58. H. Köstlin, Festkörperprobleme 22, 229 (1982).
59. P.M. Berning, Appl. Opt. 22, 4127 (1983).
60. O.S. Heavens, *Optical Properties of Thin Solid Films* (Dover, New York, 1965).
61. M. Born and E. Wolf, *Principles of Optics*, 6th edition (Pergamon, Oxford, 1980).
62. E.D. Palik, editor, *Handbook of Optical Constants of Solids* (Academic, New York, 1985).
63. K.L. Chopra, *Thin Film Phenomena* (McGraw-Hill, New York, 1969).
64. F. Abeles, Y. Borensztein and T. Lopez-Rios, Festkörperprobleme 24, 93 (1984).
65. S. Norrman, T. Andersson, C.G. Granqvist and O. Hunderi, Phys. Rev. B 18, 674 (1978).

66. G.B. Smith, G.A. Niklasson, J.S.E.M. Svensson and C.G. Granqvist, J. Appl. Phys. 59, 571 (1986).
67. P. Gadenne, A. Beghdadi and J. Lafait, Opt. Commun. 65, 17 (1988).
68. Y. Yagil and G. Deutscher, Appl. Phys. Lett. 52, 373 (1988).
69. M. Gadenne, J. Lafait and P. Gadenne, Opt. Commun. 71, 273 (1989).
70. P. Gadenne, Y. Yagil and G. Deutscher, J. Appl. Phys. 66, 3019 (1989).
71. Y. Yagil, M. Yosefin, D.J. Bergman, G. Deutscher and P. Gadenne, to be published.
72. E. Valkonen and C.-G. Ribbing, Mater. Lett. 3, 29 (1984).
73. T. Karlsson, C.-G. Ribbing, A. Roos and E. Valkonen, Int. J. Energy Res. 12, 23 (1988).
74. P.O. Jarvinen, J. Energy 2, 95 (1978).
75. I. Hamberg, J.S.E.M. Svensson, T.S. Eriksson, C.G. Granqvist, P. Arrenius and F. Norin, Appl. Opt. 26, 2131 (1987).
76. Z.M. Jarzebski and J.P. Marton, J. Electrochem. Soc. 123, 199c, 299c, 333c (1976).
77. G. Haacke, Ann. Rev. Mater. Sci. 7, 73 (1977).
78. Z.M. Jarzebski, Phys. Stat. Sol. 71, 13 (1982).
79. J.C. Manifacier, Thin Solid Films 90, 297 (1982).
80. K.L. Chopra, S. Major and K. Pandya, Thin Solid Films 102, 1 (1983).
81. I. Hamberg and C.G. Granqvist, J. Appl. Phys. 60, R123 (1986).
82. Z.-C. Jin, I. Hamberg and C.G. Granqvist, Appl. Phys. Lett. 51, 149 (1987).
83. Z.-C. Jin, I. Hamberg and C.G. Granqvist, J. Appl. Phys. 64, 5117 (1988).
84. H. Haitjema, J.J.P. Elich and C.J. Hoogendoorn, Solar Energy Mater. 18, 283 (1989).
85. B. Stjerna and C.G. Granqvist, Appl. Phys. Lett. 57, 1989 (1990).
86. M. Marezio, Acta Crystallogr. 20, 273 (1966).
87. S.-J. Jiang and C.G. Granqvist, Proc. Soc. Photo-Opt. Instrum. Engr. 562, 129 (1985).
88. Y. Sawada and Y. Taga, Thin Solid Films 116, L55 (1984).
89. H. Schade and Z.E. Smith, Appl. Opt. 24, 3221 (1985).
90. J.G. O'Dowd, Proc. Soc. Photo-Opt. Instrum. Engr. 692, 58 (1986).
91. I. Hamberg, C.G. Granqvist, K.-F. Berggren, B.E. Sernelius and L. Engström, Phys. Rev. B 30, 3240 (1984).
92. B.E. Sernelius, K.-F. Berggren, Z.-C. Jin, I. Hamberg and C.G. Granqvist, Phys. Rev. B 37, 10244 (1988).
93. C.A. Hogarth and Z.T. Al-Dhhan, Phys. Stat. Sol. 137, K157 (1986).
94. M.A. Sainz, A. Durán and J.M. Fernández Navarro, J. Non-Cryst. Solids 121, 315 (1990).
95. B.W. Faughnan and R.S. Crandall, in *Display Devices*, edited by J.I. Pankove, *Topics in Applied Physics*, Vol. 40 (Springer, Berlin, 1980), p. 181.
96. G. Beni and J.L. Shay, Adv. Image Pickup and Display 5, 83 (1982).
97. W.C. Dautremont-Smith, Displays, January 3, April 67 (1982).
98. S.A. Agnihotry, K.K. Saini and S. Chandra, Indian J. Pure Appl. Phys. 24, 19, 34 (1986).
99. S.K. Deb, Proc. Soc. Photo-Opt. Instrum Engr. 692, 19 (1986).
100. T. Oi, Ann. Rev. Mater. Sci. 16, 185 (1986).
101. A. Donnadieu, Mater. Sci. Engr. B 3, 185 (1989).
102. A.M. Andersson, C.G. Granqvist and J.R. Stevens, Appl. Opt. 28, 3295 (1989).

103. R.B. Goldner, R.L. Chapman, G. Foley, E.L. Goldner, T. Haas, P. Norton, G. Seward and K.K. Wong, Solar Energy Mater. 14, 195 (1986).
104. R.B. Goldner, G. Seward, K. Wong, G. Berera, T. Haas and P. Norton, Proc. Soc. Photo-Opt. Instrum. Engr. 823, 101 (1987).
105. G. Seward, R.B. Goldner, K. Wong, T. Haas, G.H. Foley, R. Chapman and S. Shultz, Proc. Soc. Photo-Opt. Instrum. Engr. 823, 90 (1987).
106. J.L. Lagzdons, G.E. Bajars and A.R. Lusis, Phys. Stat. Sol. A 84, K197 (1984).
107. C.M. Lampert, T.R. Omstead and P.C. Yu, Solar Energy Mater. 14, 161 (1986).
108. J.S.E.M. Svensson and C.G. Granqvist, Appl. Phys. Lett. 49, 1566 (1986).
109. M.K. Carpenter, R.S. Conell and D.A. Corrigan, Solar Energy Mater. 16, 333 (1987).
110. M. Fantini and A. Gorenstein, Solar Energy Mater. 16, 487 (1987).
111. J.S.E.M. Svensson and C.G. Granqvist, Solar Energy Mater. 16, 19 (1987).
112. J.S.E.M. Svensson and C.G. Granqvist, Appl. Opt. 26, 1554 (1987).
113. P.C. Yu, G. Nazri and C.M. Lampert, Solar Energy Mater. 16, 1 (1987).
114. K. Bange, F.G.K. Baucke and B. Metz, Proc. Soc. Photo-Opt. Instrum. Engr. 1016, 170 (1988).
115. P. Delichere, S. Joiret, A. Hugot-LeGoff, K. Bange and B. Metz, Proc. Soc. Photo-Opt. Instrum. Engr. 1016, 165 (1988).
116. W. Estrada, A.M. Andersson and C.G. Granqvist, J. Appl. Phys. 64, 3678 (1988).
117. N.R. Lynam and H.R. Habibi, Proc. Soc. Photo-Opt. Instrum. Engr. 1016, 63 (1988).
118. A. Pennisi and C.M. Lampert, Proc. Soc. Photo-Opt. Instrum. Engr. 1016, 131 (1988).
119. S. Yamada, T. Yoshioka, M. Miyashita, K. Urabe and M. Kitao, Proc. Soc. Photo-Opt. Instrum. Engr. 1016, 34 (1988).
120. P.C. Yu and C.M. Lampert, Solar Energy Mater. 19, 1 (1989).
121. S.F. Cogan, T.D. Plante, R.S. McFadden and R.D. Rauh, Proc. Soc. Photo-Opt. Instrum. Engr. 823, 106 (1987).
122. P.V. Ashrit, F.E. Girouard, V.-V. Truong and G. Bader, Proc. Soc. Photo-Opt. Instrum. Engr. 562, 53 (1985).
123. B.M. Armand, Ann. Rev. Mater. Sci. 16, 245 (1986).
124. J.R. Stevens, J.S.E.M. Svensson, C.G. Granqvist and R. Spindler, Appl. Opt. 26, 3489 (1987).
125. A. Deneuville, P. Gerard and R. Billat, Thin Solid Films 70, 203 (1980).
126. T. Yoshimura, M. Watanabe, Y. Koike, K. Kiyota and M. Tanaka, Thin Solid Films 101, 141 (1983).
127. J.S.E.M. Svensson and C.G. Granqvist, Appl. Phys. Lett. 45, 828 (1984).
128. D.K. Benson, C.E. Tracy, J.S.E.M. Svensson and B.E. Liebert, Proc. Soc. Photo-Opt. Instrum. Engr. 823, 72 (1987).
129. Y. Kamimori, J. Nagai and M. Mizuhashi, Solar Energy Mater. 16, 27 (1987).
130. R. Viennet, J.-P. Randin and I.D. Raistrick, J. Electrochem. Soc. 129, 2451 (1982).
131. S. Takaki, K. Matsumoto and K. Suzuki, Appl. Surface Sci. 33/34, 919 (1988).
132. J.P. Randin, in *Proc. First European Display Research Conf., Eurodisplay '81* (VDE-Verlag, Berlin, 1981), p. 94.
133. D. Craigen, A. Mackintosh, J. Hickman and K. Colbow, J. Electrochem. Soc. 133, 1529 (1986).

134. D. Davazoglou, A. Donnadieu, R. Fourcade, A. Hugot-LeGoff, P. Delichere and A. Perez, Rev. Phys. Appl. 23, 265 (1988).
135. H. Unuma, K. Tonooka, Y. Suzuki, T. Furusaki, K. Kodaira and T. Matsushita, J. Mater. Sci. Lett. 5, 1248 (1986).
136. N.R. Lynam, F.H. Moser and B.P. Hichwa, Proc. Soc. Photo-Opt. Instrum. Engr. 823, 130 (1987).
137. K. Sone and Y. Fukuda, *Inorganic Thermochromism*, Springer Series on Inorganic Chemistry Concepts, Vol. 10 (Springer, Berlin, 1987).
138. J.B. Goodenough, Progr. Solid State Chem. 5, 145 (1971).
139. C.B. Greenberg, Thin Solid Films 110, 73 (1983).
140. G.V. Jorgenson and J.C. Lee, Solar Energy Mater. 14, 205 (1986).
141. J.C. Lee, G.V. Jorgenson and R.J. Lin, Proc. Soc. Photo-Opt. Instrum. Engr. 692, 2 (1986).
142. S.M. Babulanam, W. Estrada, M.O. Hakim, S. Yatsuya, A.M. Andersson, J.R. Stevens, J.S.E.M. Svensson and C.G. Granqvist, Proc. Soc. Photo-Opt. Instrum. Engr. 823, 64 (1987).
143. K.A. Khan, G.A. Niklasson and C.G. Granqvist, J. Appl. Phys. 64, 3327 (1988).
144. K.A. Khan and C.G. Granqvist, Appl. Phys. Lett. 55, 4 (1989).
145. G. Mbise, G.B. Smith, G.A. Niklasson and C.G. Granqvist, Appl. Phys. Lett. 54, 987 (1989).
146. S.M. Machaggah, R.T. Kivaisi and E.M. Lushiku, Solar Energy Mater. 19, 315 (1989).
147. T. Motohiro, H. Yamadera and Y. Taga, Rev. Sci. Instrum. 60, 2657 (1989).
148. H.J. Leamy, G.H. Gilmer and A.G. Dirks, in *Current Topics in Materials Science*, edited by E. Kaldis (North-Holland, Amsterdam, 1980), Vol. 6, p. 309.
149. J. Krug and P. Meakin, Phys. Rev. A 40, 2064 (1989).
150. P. Meakin and J. Krug, Europhys. Lett. 11, 7 (1990).
151. S.-J. Jiang, Z.-C. Jin and C.G. Granqvist, Appl. Opt. 27, 2847 (1988).
152. I. Hamberg and C.G. Granqvist, Appl. Opt. 22, 609 (1983).
153. C. Amra, G. Albrand and P. Roche, Appl. Opt. 25, 2695 (1986).
154. F.H. Nicoll, RCA Rev. 10, 440 (1949).
155. S.M. Thomsen, RCA Rev. 12, 143 (1951).
156. K.J. Cathro, D.C. Constable and T. Solaga, Solar Energy 27, 491 (1981).
157. L. Ignberg, Uppsala University Technical Report UPTEC 8260E (in Swedish; unpublished, 1982).
158. H.J. Gläser, Glastechn. Ber. 50, 248 (1977).
159. D. Arasteh, S. Selkowitz and J. Hartmann, in *Proc. ASHRAE/DOE/BTECC Conference on Thermal Performance of the Exterior Envelopes of Buildings III*, Clearwater Beach, Florida, USA, 2-5 December 1985; also Lawrence Berkeley Laboratory Report LBL-20348.
160. P. Derner, Glastechn. Ber. 48, 84 (1975).
161. T.S. Eriksson, C.G. Granqvist and J. Karlsson, Solar Energy Mater. 16, 243 (1987).
162. W.J. Platzer, Solar Energy Mater. 16, 275 (1987).
163. A. Pflüger, Ph.D. Thesis, Fakultät für Physik der Albert-Ludwigs-Universität, Freiburg, Germany (unpublished, 1988).
164. W.J. Platzer, Ph.D. Thesis, Fakultät für Physik der Albert-Ludwigs-Universität, Freiburg, Germany (unpublished, 1988).

165. W.J. Platzer and V. Wittwer, in *Proc. Workshop on Optical Measurement Techniques*, Ispra, Italy, 27-29 October (1987).
166. C.A. Bishop and R.P. Howson, Solar Energy Mater. 13, 175 (1986).
167. K. Chiba, S. Sobajima and Y. Yatabe, Solar Energy Mater. 8, 371 (1983).
168. R.P. Howson, M.I. Ridge and K. Suzuki, Proc. Soc. Photo-Opt. Instrum. Engr. 428, 14 (1983).
169. Z.-C. Jin and C.G. Granqvist, Appl. Opt. 26, 3191 (1987).
170. T. Minami, H. Nanto and S. Takata, Japan. J. Appl. Phys. 27, L287 (1988).
171. B. Stjerna and C.G. Granqvist, Solar Energy Mater. 20, 225 (1990); Appl. Opt. 29, 447 (1990).
172. A. Pflüger, W.J. Platzer and V. Wittwer, in *Advances in Solar Energy Technology*, edited by W.H. Bloss and F. Pfisterer (Pergamon, Oxford, 1988), p. 636.
173. W.J. Platzer and V. Wittwer, *Transparente Wärmedämmaterialen für den Einsatz im Solarenergiebereich* (1988).
174. V. Wittwer, private communication.
175. A. Pflüger, Solar Energy Mater. 16, 255 (1987).
176. J. Fricke, J. Non-Cryst. Solids 100, 169 (1988).
177. J. Fricke, Sci. Am. 258, (5), 92 (1988).
178. J. Fricke, editor, *Aerogels*, Springer Proc. in Physics, Vol. 6 (Springer, Berlin, 1986).
179. S. Henning and L. Svensson, Phys. Scripta 23, 697 (1981).
180. S. Henning, in *Aerogels*, edited by J. Fricke (Springer, Berlin, 1986), p. 38.
181. M. Rubin and C.M. Lampert, Solar Energy Mater. 7, 393 (1983).
182. W.J. Platzer, V. Wittwer and M. Mielke, in *Aerogels*, edited by J. Fricke (Springer, Berlin, 1986), p. 127.
183. P.H. Tewari, A.J. Hunt, J.G. Lieber and K. Lofftus, in *Aerogels*, edited by J. Fricke (Springer, Berlin, 1986), p. 142.
184. J. Fricke, R. Caps, D. Büttner, U. Heinemann, E. Hümmer and A. Kadur, Solar Energy Mater. 6, 267 (1987).
185. E. Boy, Bauphysik 11, 21 (1989).
186. D.K. Benson, C.E. Tracy and G.J. Jorgensen, Proc. Soc. Photo-Opt. Instrum. Engr. 502, 146 (1984).
187. S. Robinson and R.E. Collins, in *Proc. ISES Solar Energy World Congress*, Kobe, Japan, 1990, to be published.
188. D.K. Benson, C.E. Tracy, T. Potter, C. Christensen and D.E. Soule, Solar Energy Research Institute Report SERI SP-255-3318 (1988).
189. N.Y.C. Chu, Solar Energy Mater. 14, 215 (1986).
190. T. Novinson, Proc. Soc. Photo-Opt. Instrum. Engr. 823, 138 (1987).
191. G. Reusch, Glaswelt 21, (1), 16 (1968).
192. J. Germer, Solar Age, October 1984, p. 20.
193. E. Boy and S. Meinhardt, in *Proc. Second International Workshop on Transparent Insulation Materials in Solar Energy Conversion for Buildings and Other Applications*, Freiburg, Germany, 24-25 March (1988).
194. A.M. Andersson, G.A. Niklasson and C.G. Granqvist, Appl. Opt. 26, 2164 (1987).
195. Y. Anjaneyulu and D.W. Yoon, Solar Energy Mater. 14, 223 (1986).
196. P. van Konynenburg, S. Marsland and J. McCoy, Solar Energy Mater. 19, 27 (1989).

Chapter 6

MATERIALS FOR RADIATIVE COOLING TO LOW TEMPERATURES

C.G. Granqvist and T.S. Eriksson

Physics Department
Chalmers University of Technology and University of Gothenburg
S-412 96 Gothenburg, Sweden

ABSTRACT

Radiative cooling uses the clear sky as a heat sink. To assess the potential of this free and abundant source of cooling we first present computed data, based on a detailed model of the sky radiance, which show that a cooling power of ~ 100 Wm^{-2} at ambient temperature and a maximum practical temperature difference of ~ 25°C can be accomplished. Low temperature applications hinge on materials development. We review work on selectively infrared-emitting surfaces with a focus on silicon-based coatings backed by metal, metallized polymer foils, gas slabs backed by metal, and certain ceramic oxide layers. We also treat infrared-transparent polyethylene-based convection shields with a focus on cellular constructions, as well as coatings and pigments for diminishing the solar transmittance. The results of a few selected field tests are included. Under favourable, but not uncommon, meteorological conditions one can reach temperature differences of 15-20°C during the night and ~ 10°C during the day with simple devices.

I. INTRODUCTION

The clear sky can serve as a heat sink and can thus be used to produce temperatures below those of the ambience in a purely passive way. This property of the clear sky has been known since ancient times, and scientific studies of attainable temperature differences date back at least as far as to the first part of the 19th century.[1] Cooling under clear weather conditions has numerous important practical consequences related to temperature control of the Earth, meteorology, natural climatization, etc. The consequences to Man can be both good and bad: As an example where the cooling causes problems, we can mention that growing crops can be damaged by frost even if the air temperature is several degrees above 0°C, as is well known to farmers and gardeners. Another example, where natural cooling is used to advantage, can be found in traditional Iranian ice-makers and desalination ponds.[2]

Even if passive cooling under clear skies is a widely recognized phenomenon, the development of materials and devices for efficient utilization of this free and abundant source of cooling is a relatively new subject.[3-6] The key to a conscious materials development lies in an understanding of the balance between the radiation which wells down from the atmosphere, and the radiation emitted from a surface exposed towards the atmosphere. Either type of radiation may be characterized by a strong spectral dependence. Section II introduces the subject by presenting computed radiance spectra for different model atmospheres and giving estimates of cooling power (~ 100 Wm^{-2} at ambient temperature) and maximum temperature drop (< 25°C in a practical device). Low-temperature applications of radiative cooling - as we henceforth call the phenomenon of passive cooling under clear skies - hinge on materials development. Specifically, two materials issues are of central importance. They are surface treatments giving infrared selectivity with high emittance in the 8-13-μm wavelength interval and low absorptance elsewhere, and convection shields with significant infrared transmittance. Section III reviews work on infrared-selective surfaces, with discussion of silicon-based coatings backed by metal, metallized polymer foils, gas slabs backed by metal, and certain ceramic oxide layers. Section IV summarizes results on infrared-transparent convection shields, for which polyethylene-based materials have been used without exception. We treat cellular materials as well as foils with significant solar reflectance. Section V is devoted to results from a few selected field tests. We include data obtained using selectively infrared-emitting surfaces placed under transparent convection shields, certain innovative multistage cooling devices, and polyethylene-based solar reflecting foils.

Based on the work conducted so far, one can state that there is no doubt regarding the theoretical assessments of the potential of radiative cooling, or that such cooling can be applied in practice. Under favourable - though not uncommon - meteorological conditions one can reach temperature differences of 15-20°C during the night and ~ 10°C during the day by use of simple single-stage devices. Multistage devices can reach even lower temperatures. The lowest temperatures require selectively infrared-emitting surfaces, but if premium is put on a large cooling power at a moderate temperature difference, a blackbody-like radiator is often the best option. Solar-reflecting convection shields appear to offer a cheap means for moderate space cooling even during the day.

II. THE RESOURCE FOR RADIATIVE COOLING

This section presents data on the spectral radiance from different model atmospheres representative of various latitudes (Sec. II A), as well as on the relationship between cooling power and temperature difference (Sec. II B).

A. Atmospheric Radiance

The radiance coming from the atmosphere is extremely complicated.[7,8] The main gases, nitrogen and oxygen, contribute very little, whereas the variable constituents like water vapour, carbon dioxide, ozone, and - to a much smaller extent - nitrogen oxides and hydrocarbons show important absorption bands in the thermal infrared range. This is taken to be $3 < \lambda < 100$ µm, where λ denotes the wavelength. Water vapour has a strong split vibrational band centered around 6.3 µm and also shows significant rotational absorption at $\lambda > 20$ µm.[9] This absorption can extend to shorter wavelengths if the humidity is high. Carbon dioxide has a broad intense vibrational band centered at ~ 15 µm.[9] Ozone absorption is not equally important but several absorption bands lie in the infrared.[9] Most of these are masked by water vapour and carbon dioxide, but a narrow absorption band at 9.6 µm shows up distinctly. One concludes that the atmospheric radiation downward shows two important features: First, the spectral radiance has a minimum in an interval which lies between the major absorption bands of water vapour and carbon dioxide. Second, the overall spectral radiance depends strongly on climatic conditions - particularly on the amount of water vapour. These features have been verified numerous times by direct spectroradiometric measurements (see, for example, Refs. 10-12).

For quantitative assessments of the cooling resource, it is useful to start from detailed data on the spectral sky radiance representative of typical climates at different latitudes. Such results can be extracted from a computer program known as[13] LOWTRAN 5 (or a more recent version of this). It uses a single-parameter band model for the molecular absorption and includes the effects of continuum absorption, molecular scattering, and aerosol extinction. Atmospheric refraction and earth curvature are included for slant atmospheric paths. For radiative cooling purposes, the LOWTRAN 5 program needs to be used exclusively in the radiance mode, in which a numerical evaluation of the integral form of the radiative transfer equation is employed. The emission from aerosols and the treatment of aerosol and molecular scattering are considered only to zeroth order; additional contributions to atmospheric emission from scattered radiation are neglected. Local thermodynamic equilibrium is assumed. The radiance data obtained from LOWTRAN are known to be in very good overall agreement with measured results.[13]

The LOWTRAN 5 program is provided with data for five seasonal atmospheres together with the 1962 U.S. standard atmosphere. It is also possible to replace these by user derived or measured values. The seasonal models are representative of the following atmospheres: tropical (15° N), midlatitude summer (45° N, July), midlatitude winter (45° N, January), subarctic summer (60° N, July) and subarctic winter (60° N, January). The atmospheres are specified in terms of height profiles for temperature, barometric pressure, and densities of water vapour, ozone, nitric acid, and of the uniformly mixed gases (CO_2, N_2O, CH_4, CO, N_2 and O_2). The LOWTRAN 5 program also contains several aerosol models. Figure 1 shows atmospheric zenith radiance for all six model atmospheres. We denote this quantity by $L_a\ (\theta = 0, \lambda, \tau_a)$, with τ_a being the temperature of the atmospheric boundary layer. The data are reproduced from Ref. 14. These

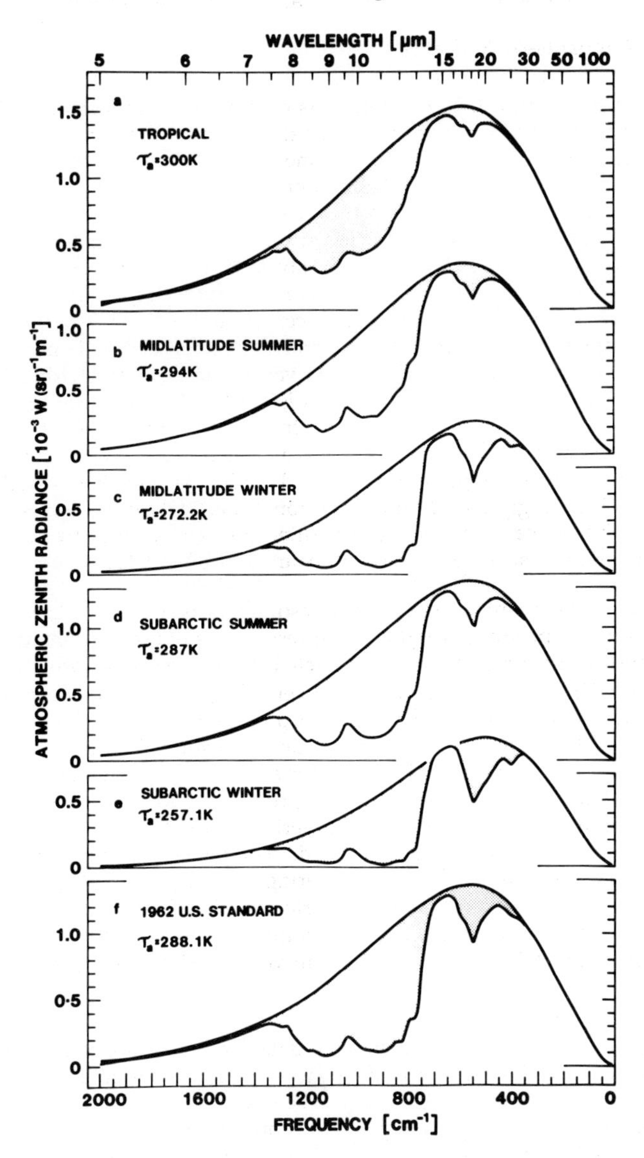

Fig. 1. Calculated spectral zenith radiance (lower curves in parts a-f) and corresponding blackbody radiance (upper curves in parts a-f) for six model atmospheres, as computed from the LOWTRAN 5 model. The horizontal axis shows frequency as well as wavelength. (From Ref. 14).

curves are seen to drop below, or to approximately follow, blackbody radiance spectra defined by τ_a, denoted L_{bb} (λ, τ_a), for frequencies above 350 cm^{-1}. At lower frequencies, where no LOWTRAN data are available, only the blackbody curve is plotted. We find for all atmospheres that the actual radiance lies far below that of the blackbody in the 8-13-μm range (known as the "atmospheric window"). A secondary window at 16-22 μm is of much smaller significance.

The angular dependence of the atmospheric radiance is of importance for evaluations of the radiative cooling resource. Figure 2 shows this property for one particular model atmosphere, the 1962 U.S. standard, at four different zenith angles.[14] It is seen that the radiance is enhanced within the "atmospheric windows" when the zenith angle θ is increased (because the path length contributing to the radiation goes up)[5] and that the various curves practically overlap in those spectral ranges where the emission from water vapour and carbon dioxide are strongest. The solid curve refers to θ = 45° and is hence representative of the hemispherical radiance. By comparison with the bottom curve in Fig. 1, it is found that the radiance at θ = 45° is only slightly higher than the radiance from the zenith direction. For larger zenith angles the "atmospheric window" gradually becomes closed. The atmospheric hemispherical radiance, which is pertinent to assessments of radiative cooling for surfaces freely exposed to the skies, can be obtained with accuracy by integrating over θ.

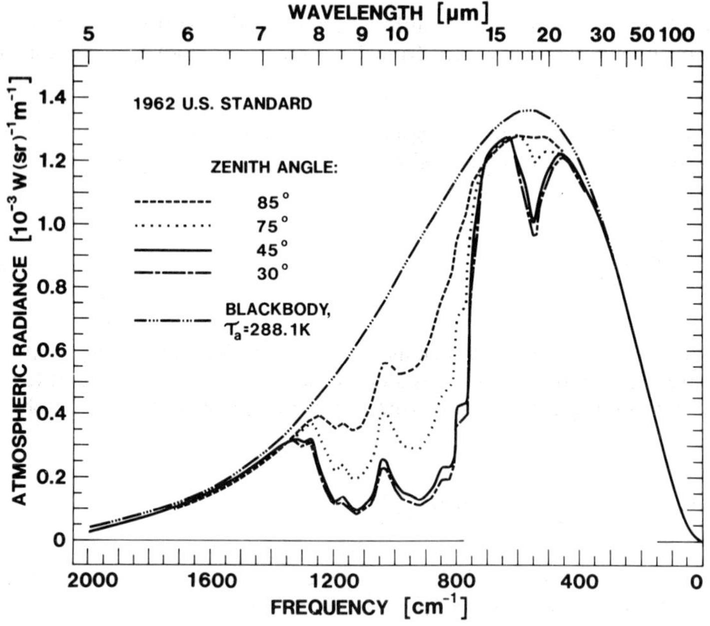

Fig. 2. Calculated spectral radiance for the 1962 U.S. standard atmosphere at four zenith angles. A blackbody curve corresponding to 288.1 K ambient temperature is included for comparison. The horizontal axis shows frequency as well as wavelength. (From Ref. 14).

The data in Figs. 1 and 2 represent cloud-free atmospheres. If the amount of precipitable water is increased significantly, the "atmospheric window" is much less apparent; detailed results are given in Refs. 5 and 11. A thick and low-lying cloud-cover eliminates radiative cooling entirely. Thin and high-lying clouds are less significant; their influence can be estimated only through calculations based on actual height profiles for water vapour density and temperature.

It is convenient to introduce an atmospheric emittance $e_a(\theta, \lambda)$ defined by

$$e_a(\theta, \lambda) = L_a(\theta, \lambda, \tau_a) \;/\; L_{bb}(\lambda, \tau_a). \tag{1}$$

Clearly, $e_a(\theta, \lambda)$ is significantly different from unity only in the 8-13 μm range, and one can define a "box model"[5] by

$$e_a(\theta, \lambda) = 1, \quad \text{for } 3 < \lambda < 8\ \mu\text{m}, \tag{2}$$

$$e_a(\theta, \lambda) = 1 - [1 - \bar{e}_{a2}(0)]^{1/\cos\theta}, \quad \text{for } 8 < \lambda < 13\ \mu\text{m}, \tag{3}$$

$$e_a(\theta, \lambda) = 1, \quad \text{for } \lambda > 13\ \mu\text{m}. \tag{4}$$

Here $\bar{e}_{a2}(0)$ denotes an average zenith emittance in the "atmospheric window", which in principle can be obtained from LOWTRAN or from spectroradiometry with suitable filters.

Some simplified formulas for hemispherically averaged emittance values are known from the literature. These relate the integrated quantity, expressed generally as

$$\bar{x}^H \equiv \int_0^{\pi/2} d(\sin^2\theta)\, x(\theta), \tag{5}$$

to readily accessible parameters such as water vapour density or dew point temperature τ_{dp}. Recent work[15] has shown that hemispherical thermal (i.e., integrated over the full Planck spectrum) atmospheric emittance can be obtained from

$$\bar{e}_a^H = 0.711 + 0.56\left(\frac{\tau_{dp}}{100}\right) + 0.73\left(\frac{\tau_{dp}}{100}\right)^2, \tag{6}$$

where τ_{dp} is in °C. This formula is useful for assessing radiative cooling of black-body-like surfaces. A relation for $\bar{e}_{a2}^H$ is given in Ref. 16.

B. Cooling Power and Temperature Difference for Ideal Surfaces

The LOWTRAN data of $L_a(\theta, \lambda, \tau_a)$ can be used for quantitative predictions of cooling power and achievable temperature difference. To this end we consider a surface which faces the sky and derive the net thermal radiative flux as the difference between outgoing and incoming contributions according to[5]

$$P_{rad} = \pi \int_0^{\pi/2} d(\sin^2 \theta) \int_0^{\infty} d\lambda \, [1 - R(\theta, \lambda)] \, [L_{bb}(\lambda, \tau_s) - L_a(\theta, \lambda, \tau_a)]. \tag{7}$$

Here τ_s denotes the temperature of the exposed (non-transparent) surface and $1 - R$ is its absorptance or, equivalently, emittance. The off-normal reflectance must be regarded as the arithmetic mean of the reflectance due to TE- and TM-polarised radiation, i.e.,

$$R(\theta, \lambda) = \frac{1}{2} [R_{TE}(\theta, \lambda) + R_{TM}(\theta, \lambda)]. \tag{8}$$

Radiative cooling causes a temperature drop ΔT which is given by

$$\Delta T = \tau_a - \tau_s. \tag{9}$$

At $\Delta T > 0$ it is necessary to regard the role of a non-radiative heat influx to the exposed surface. This limits the practically useful cooling power P_c to

$$P_c = P_{rad} - \kappa \Delta T, \tag{10}$$

where the loss is specified in terms of a linear heat-transfer coefficient κ.

It is evident from Eq. (7) that the spectral surface reflectance governs the radiative cooling. It is illustrative to consider three types of idealized surfaces. The first of these is the fully reflecting surface with $R(\theta, \lambda) = 1$. It serves as an approximation for good metallic surfaces (coated with $<< 0.1$ μm of oxide, etc.) which can have $R \approx 0.99$ in the thermal infrared. Such surfaces experience no noticeable radiative cooling, as observed already in the very first studies[1] on this subject. We remark, in passing, that transparent and infrared-reflecting SnO_2:F coatings can prevent radiative cooling of glass surfaces exposed to the clear sky. Such (electrically conducting) coatings have been used to eliminate the formation of thick frost layers, which otherwise occur on the windscreens of cars parked outdoors during clear nights. Those aspects of radiative cooling are treated further in Ref. 17. Similarly, surface coatings based on non-conducting BeO can prevent radiative cooling and frost formation;[18] they may be of interest for high-voltage power lines.

Our second sample regards a blackbody-radiating surface defined by $R(\theta, \lambda) = 0$. This surface yields the largest cooling power at ambient temperature. A blackbody serves as a good approximation for organic matter, soils, rocks, water, ice, most paint layers, concrete, asphalt, ordinary uncoated glass, etc.

The third, and for our present purposes most interesting, example is the selectively infrared-emitting surface designed for reaching the lowest possible temperature. This surface should have high emittance in the 8-13 μm "atmospheric window" range, where the counter radiation is weak, and low absorptance outside this interval, so that the main portion of the radiation from H_2O and CO_2 is not interacting with the surface. Hence the ideal property is

$$R(\theta, \lambda) = R_{sel}(\theta, \lambda) = 0, \text{ for } 8 \leq \lambda \leq 13 \text{ μm}, \quad (11)$$

$$= 1, \text{ elsewhere}. \quad (12)$$

In Sec. III below we discuss practical surfaces which approximate this ideal property.

Figure 3 shows calculated results of P_{rad} as a function of ΔT. The data apply to surfaces which radiate freely toward model atmospheres of the six types earlier discussed. In Fig. 3a the radiating surface is taken to be a blackbody; in Fig. 3b it has an ideal infrared-selective characteristic according to Eqs. (11) and (12). For both cases, the incoming power is governed by the hemispherical radiance. It is found that the radiative cooling power at ambient temperature lies between 71 and 113 Wm^{-2} for the blackbody surface and between 58 and 93 Wm^{-2} for the infrared-selective surface. The highest cooling powers hold for the U.S. STD atmosphere and the lowest for the TROP atmosphere. The values for the infrared-selective surface lie below those for the blackbody surface since only the latter takes advantage of the nonzero magnitude of $(L_{bb} - L_a)$ outside the 8-13 μm interval (cf. Fig. 1).

The radiative cooling power is seen to decrease monotonically with increasing ΔT. The decrease is much slower for the infrared-selective surface than for the blackbody surface, since the former employs radiation balance only in the 8-13 μm range where the atmospheric radiance is weak. It is seen that ultimate temperature differences between 14 and 26°C for the blackbody surface and between 27 and 62°C for the infrared-selective surface are predicted. Such large ΔT's cannot be obtained in practice, though, but the role of conductive and convective losses of cooling power must be included by use of a nonzero heat transfer coefficient. As a practical low limit, obtainable by use of an efficient infrared-transparent convection shield, we set $\kappa = 1$ $Wm^{-2}K^{-1}$. We return to this point in Sec. IV. This requirement makes the shaded triangular areas in Fig. 3 inaccessible for a cooling device. It is seen from the figure that temperature differences between 11 and 21°C for the blackbody surface and between 18 and 33°C for the infrared-selective surface are expected with $\kappa = 1$ $Wm^{-2}K^{-1}$. Again the higher values pertain to U.S. STD and the lower to TROP. Analogous computations for exchange only with the zenith atmospheric radiation are found in Ref. 13. Some improvement of the radiative cooling performance can be obtained under such conditions.

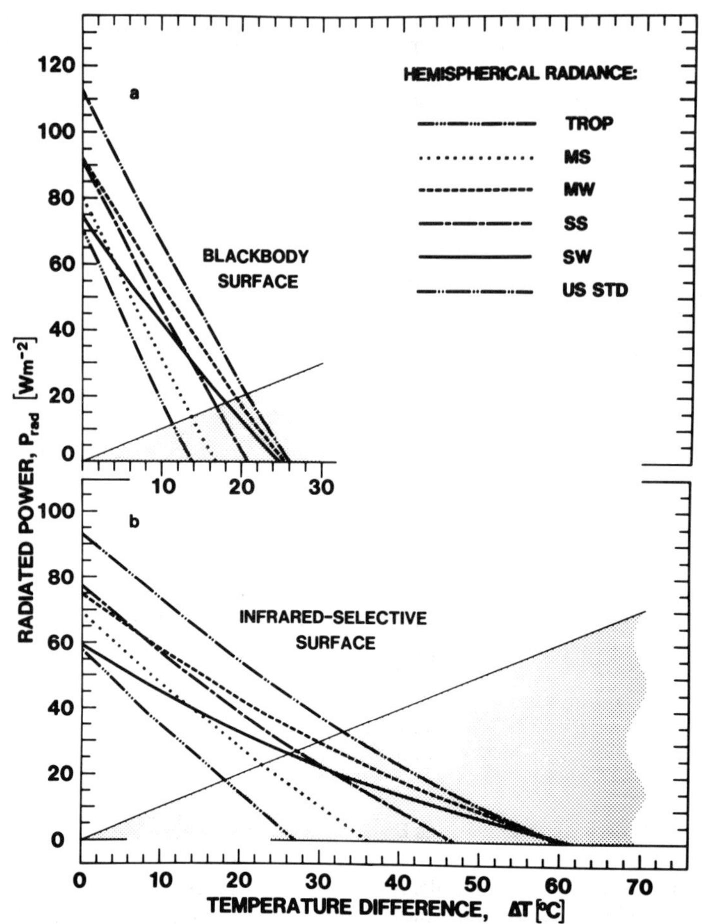

Fig. 3. Calculated relationship between radiative cooling power and temperature difference for six model atmospheres (cf. Fig. 1) and two ideal surfaces. (From Ref. 14).

III. SURFACES WITH INFRARED-SELECTIVE EMISSION

Surfaces which are infrared-selective according to Eqs. (11) and (12) have a potential for reaching lower temperatures than non-selective surfaces. Spectral selectivity can be obtained by several means: by thin silicon-based coatings backed by metal (Sec. III A), by metallized polymer foils (Sec. III B), by gas slabs backed by metal (Sec. III C), and by certain ceramic oxide layers (Sec. III D).

It is evident that none of the mentioned approaches leads to a surface which completely mimics the ideal radiative property. In order to be able to optimize a practical surface for radiative cooling applications, it is therefore convenient to define two parameters which govern the essential features of the radiative cooling. The "box model" for the atmospheric emittance, given by Eqs. (2) - (4), leads naturally to integrated surface emittance values according to[5]

$$\bar{e}_s^H = \int_0^\infty d\lambda\, L_{bb}(\lambda, \tau_a)\,[1 - R^H(\lambda)] / \int_0^\infty d\lambda\, L_{bb}(\lambda, \tau_a), \tag{13}$$

$$\bar{e}_{s2}^H = \int_{8\,\mu m}^{13\,\mu m} d\lambda\, L_{bb}(\lambda, \tau_a)\,[1 - R^H(\lambda)] / \int_{8\,\mu m}^{13\,\mu m} d\lambda\, L_{bb}(\lambda, \tau_a), \tag{14}$$

$$\eta^H = \bar{e}_{s2}^H / \bar{e}_s^H, \tag{15}$$

where R^H is the hemispherical reflectance defined, in analogy with Eq. (5), by

$$R^H(\lambda) = \int_0^{\pi/2} d(\sin^2\theta)\, R(\theta, \lambda). \tag{16}$$

Essentially, $\bar{e}_{s2}^H$ governs the cooling power at ambient temperature and η^H governs the maximum achievable temperature drop. Efficient cooling requires a large value - ideally unity - of $\bar{e}_{s2}^H$, and if low temperatures are to be reached we also require a large magnitude of η^H. The theoretical maximum of the latter quantity is 3.39 at $\tau_a = 0°C$. Several of the practical infrared-selective surfaces discussed below will be analyzed in terms of $\bar{e}_{s2}^H$ and η^H.

A. Silicon-Based Coatings Backed by Metal

Thin coatings of Si-based materials, including Si-O and Si-N bonds, have been studied in considerable detail for radiative cooling to low temperatures. The basic idea is to start with a high-reflecting metal surface, such as aluminium, and cover it with a coating which is selectively emitting in the 8-13 μm range. By choosing a proper thickness of the coating, antireflection can be used to maximize the emittance within the "atmospheric window" range. Initial work[3,5] used SiO coatings produced by resistive evaporation. This material is capable of yielding a high emittance only in part of the 8-13 μm interval, and SiO_xN_y is a superior coating material. Silicon-oxynitride coatings have been prepared for radiative cooling applications both by reactive e-beam evaporation[19] and reactive radio frequency magnetron sputtering.[20]

Figure 4a-c illustrates the infrared near-normal reflectance in the $5<\lambda<50$ μm interval for a pure silicon dioxide coating (SiO_2; part a), an essentially pure silicon nitride coating ($SiO_{0.25}N_{1.52}$; part b), and a silicon-oxynitride coating ($SiO_{1.47}N_{0.54}$; part c). All coatings are about 1 μm thick and backed by highly reflecting aluminium. The compositions were determined by Rutherford Backscattering

Spectrometry. The pronounced reflectance minima stem from phonon absorption. It appears that pure silicon dioxide yields absorption in the short-wavelength part of the "atmospheric window", and that silicon nitride yields absorption predominantly in the long-wavelength part of the "atmospheric window". Not surprisingly, a silicon-oxynitride coating, with substantial amounts of both oxygen and nitrogen, gives a good coverage of the entire "atmospheric window" range, as evident from Fig. 4c. The radiative properties of silicon-oxynitrides can be understood in detail by first assuming their microstructure to be described by five fundamental Si-based tetrahedra, whose relative occurrence is given by the stoichiometry, and then using Effective Medium Theory to average over the different tetrahedra.[20,21] The pertinent theories were introduced in the chapter on Optical Properties of Two-Component Materials. Superimposed Si-oxide/Si-nitride layers [22,23] do not produce as favourable radiative properties as those found in the silicon-oxynitrides.

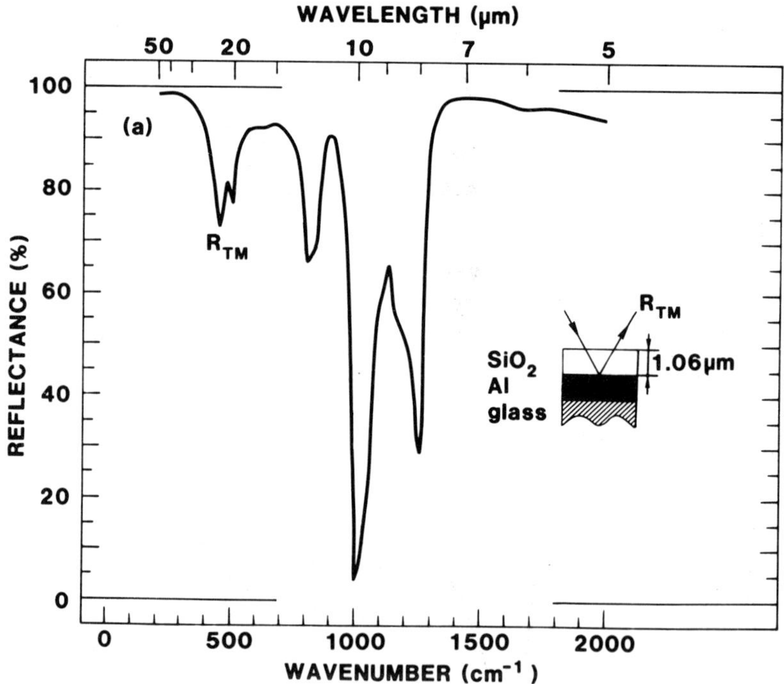

Fig. 4. Measured spectral infrared reflectance at 25° angle of incidence for reactively RF-sputtered SiO_xN_y films on aluminium coated glass. The data refer to TM polarisation, but the smallness of the angle of incidence guarantees that the reflectance of TE polarised light is practically the same. Parts a-c refer to different compositions ranging from pure silicon dioxide to almost pure silicon nitride. The insets indicate the experimental configurations and state values of film thicknesses and compositions. Part a is shown above; parts b - c follow on the next page. (Redrawn from Ref. 20).

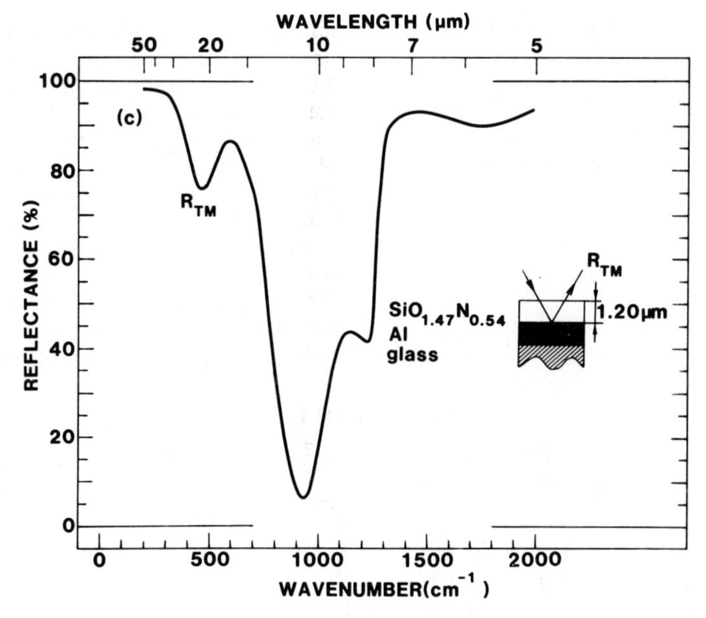
WAVELENGTH (μm)
50 20 10 7 5
100
80
60
40
20
0
(c)
R_{TM}
REFLECTANCE (%)
R_{TM}
$SiO_{1.47}N_{0.54}$
Al
glass
1.20μm
0 500 1000 1500 2000
WAVENUMBER(cm^{-1})

Fig. 4b.

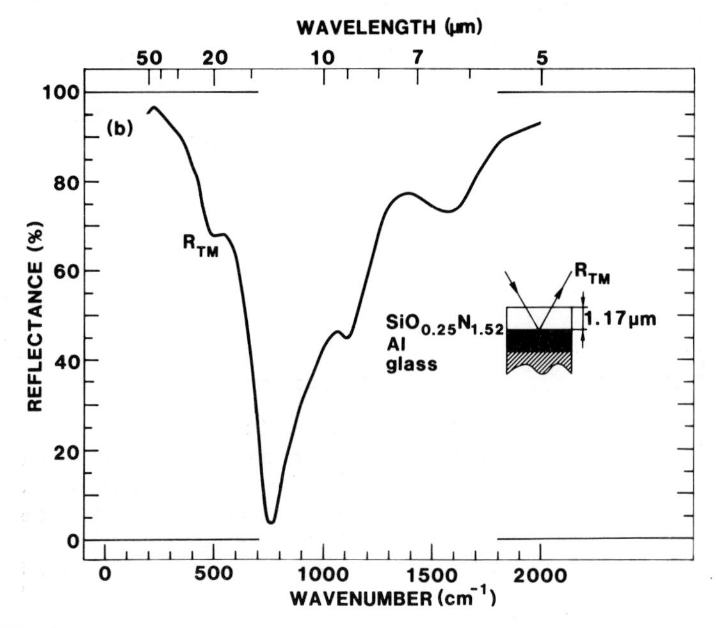
WAVELENGTH (μm)
50 20 10 7 5
100
80
60
40
20
0
(b)
R_{TM}
REFLECTANCE (%)
R_{TM}
$SiO_{0.25}N_{1.52}$
Al
glass
1.17μm
0 500 1000 1500 2000
WAVENUMBER (cm^{-1})

Fig. 4c.

Figure 5 shows spectral infrared reflectance, analogous with Fig. 4, for a $SiO_{0.6}N_{0.2}$ coating produced by evaporation of Si_3N_4 in the presence of some oxygen. The curves refer to TE and TM polarisation and 45° angle of incidence. The two sets of data are quite similar and show low reflectance selectively in the 8-13 μm range (indicated by dotted lines).

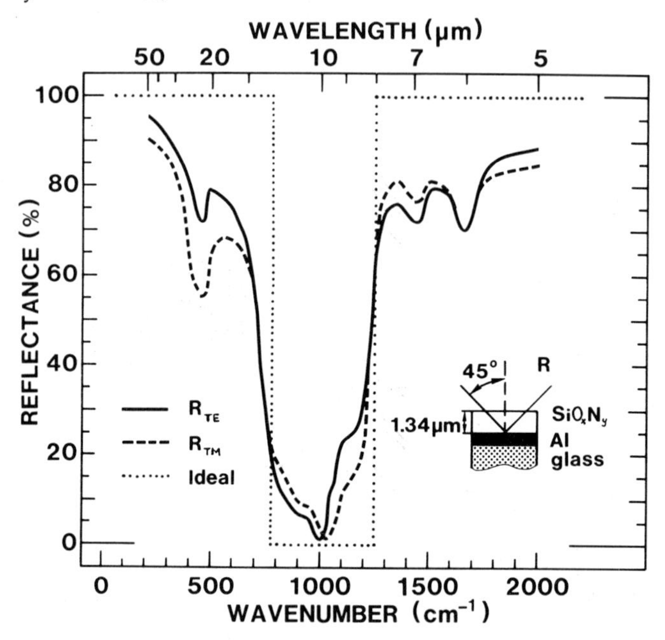

Fig. 5. Measured spectral infrared reflectance at 45° angle of incidence for an e-beam evaporated SiO_xN_y film on aluminium coated glass. The curves refer to TE and TM polarisation. Dotted lines show the ideal spectral profile for radiative cooling to low temperatures.

Spectrophotometric data on transmittance and reflectance were used to evaluate the complex dielectric functions of various silicon-based coating materials,[5,19,20] from which the integrated emittance values $\bar{e}_{s2}^{H}$ and η^{H} were derived.

Figure 6 shows results for SiO_2 and four different silicon-oxynitrides made by reactive RF-sputtering. Coatings with substantial amounts of oxygen as well as nitrogen are seen to yield large magnitudes of $\bar{e}_{s2}^{H}$ and η^{H} when the thickness is about 1 μm or somewhat higher. Figure 7 reports corresponding data for $SiO_{0.6}N_{0.2}$ and SiO made by evaporation. The oxynitride has values that are consistent with those for the corresponding sputter-deposited coatings. The SiO coating has an inferior magnitude of $\bar{e}_{s2}^{H}$. Values of $\bar{e}_{s2}^{H}$ and η^{H} for Al_2O_3 coatings are given in Ref. 24.

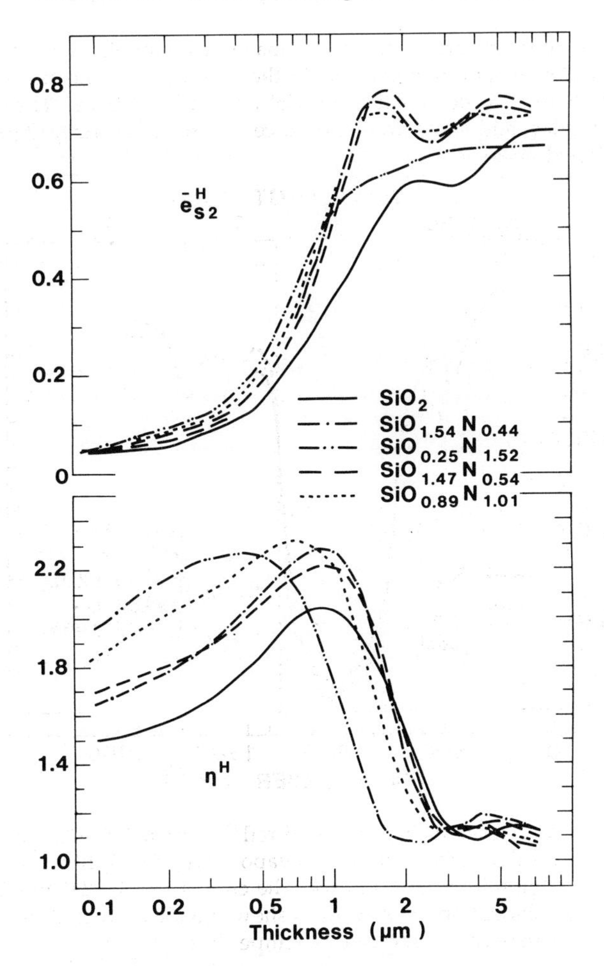

Fig. 6. Basic cooling parameters, $\bar{e}^{H}_{s2}$ and η^{H}, versus thickness for Si-based coatings made by reactive RF-magnetron sputtering.

B. Metallized Polymer Foils

As long as two decades ago it was proposed by Trombe[4] that certain metallized polymer foils could be used for radiative cooling. Some of these foils exhibit a fair degree of spectral selectivity. Figure 8 reports transmittance data for three different types of polymer foils. The shown results correspond to the expected reflectance spectrum for a foil having half the thickness stated for the transmittance curves, provided that the unexposed side of the polymer is covered with an opaque and highly reflecting metal layer (normally aluminium). The upper

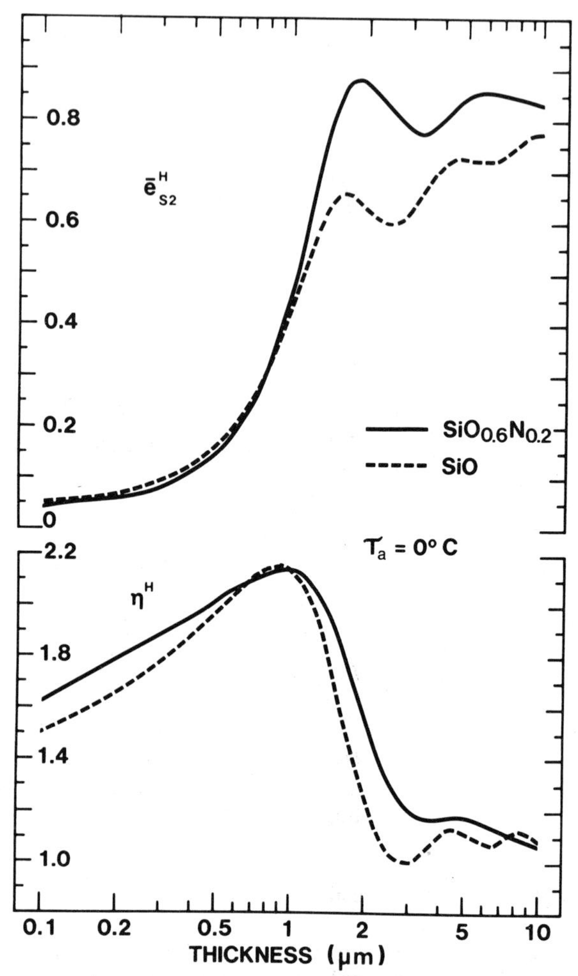

Fig. 7. Basic cooling parameters, $\bar{e}^H_{s2}$ and η^H, versus thickness for Si-based coatings made by evaporation.

curve in Fig. 8 pertains to 100 μm of polyvinylchloride (PVC). The spectral selectivity is quite weak. The other two curves apply to 12.5 μm of polyvinylfluoride (Tedlar; data from Ref. 25) and 340 μm of poly-4-methylpentene (TPX; data from Ref. 26). These latter foils are spectrally selective, although their far-infrared absorption is undesirably large. Integrated emittance values were estimated[5] from the spectral transmittance curves. It was found tht PVC yielded $\bar{e}^H_{s2} = 0.78$ and $\eta^H = 1.2$, Tedlar yielded $\bar{e}^H_{s2} = 0.84$ and $\eta^H = 1.6$, and TPX yielded $\bar{e}^H_{s2} = 0.89$ and $\eta^H = 1.6$. It should be noted that TPX appears black whereas the other foils are transparent. None of the metallized polymer foils is as selective as the best metal-backed silicon-oxynitride coatings.

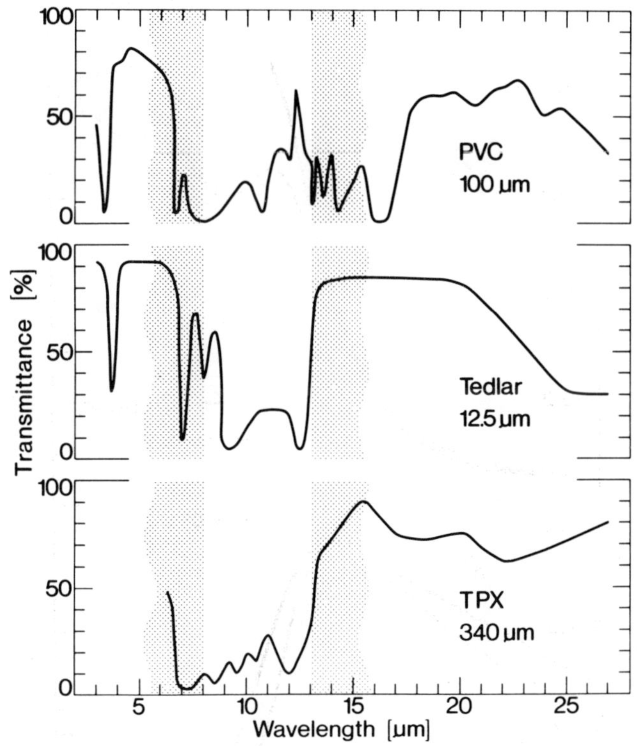

Fig. 8. Spectral transmittance for three polymer foils of interest for radiative cooling. (From Ref. 5).

C. Gas Slabs Backed by Metal

As an alternative to metal-backed silicon-based coatings or polymer foils, one may use slabs of certain selectively infrared-emitting gases confined by an infrared-transparent foil and backed by a highly reflecting metal. A thorough survey[27] of ~ 200 available organic and inorganic gases led to the conclusion that those of major interest were ammonia (NH_3), ethylene (C_2H_4), ethylene oxide (C_2H_4O), and mixtures of the latter two. The radiative properties can be related to their molecular configurations.[9,28] Gases have the advantage of being cheap, and allowing easy transport and heat-exchange of the coolant. They also permit applications which require the mixing of two or more components.

Figure 9 shows infrared transmission spectra for NH_3 with three different path lengths and at normal pressure. The gas cells were tubes with two 12 µm thick polyethylene windows. The upper graphs in Figs. 9a-c indicate the transmittance through air-filled cells and the lower graphs were recorded with gas-filled cells. The differences among the pairs of curves (shaded areas) show that NH_3 displays a

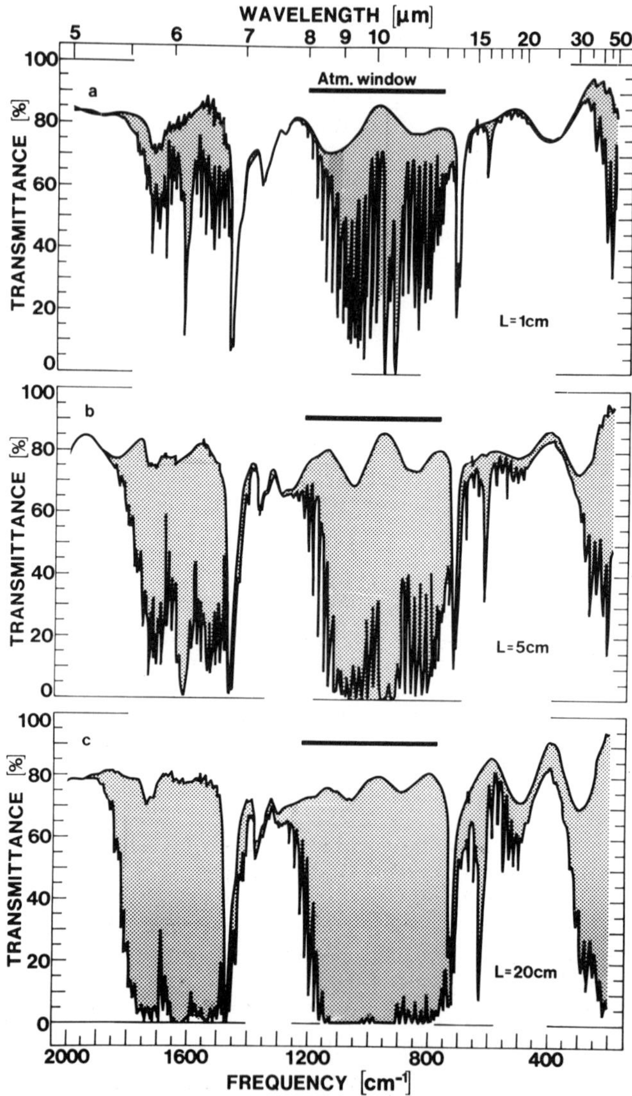

Fig. 9. Spectral transmittance for ammonia gas at normal pressure as measured with cells having the shown lengths, L. (From Ref. 27).

broad absorption band covering the desired 8-13 μm interval for path lengths exceeding ~ 1 cm. The interesting absorption is centered around an infrared-active fundamental at the frequency 950 cm^{-1}. The corresponding vibration can be visualized as the N atom moving perpendicular to the H_3 plane of the ammonia molecule while its pyramidal configuration is retained.[9] This fundamental undergoes broadening as a result of several mechanisms including inversion doubling, rotation (i.e., quantization of the angular momentum), centrifugal deformation and pressure effects. It is seen from Fig. 9 that the absorption becomes too high at the largest path lengths and, in particular, a strong absorption due to a thermal population of free rotational levels sets in at the far-infrared end of the spectrum. Transmittance spectra for C_2H_4 and C_2H_4O display a strong absorption in the 8-13 μm band, similar to the case of NH_3, and hence the two hydrocarbon gases are useful for radiative cooling. The hydrocarbons do not show any free rotational absorption in the far-infrared, which can be related to their molelcular structure.[9]

The transmittance data for the three gases were used to derive spectral absorption coefficients, from which integrated emittance values were obtained in principally the same manner as for the solid coatings and foils. Figure 10 shows $\bar{e}_{s2}^{H}$ and η^{H}

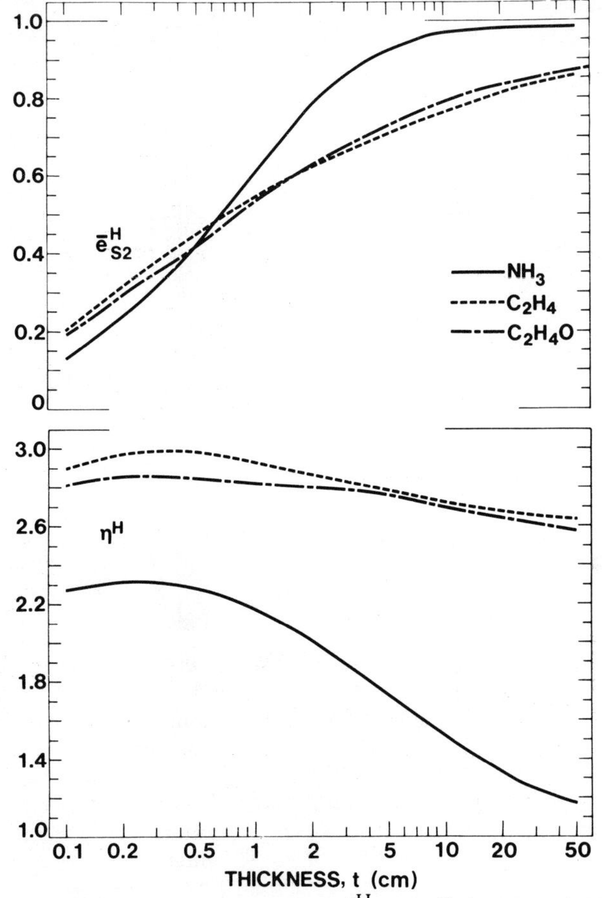

Fig. 10. Basic cooling parameters, $\bar{e}_{s2}^{H}$ and η^{H}, for NH_3, C_2H_4 and C_2H_4O gas layers having different thicknesses. (From Ref. 27).

for gas slabs of thicknesses between 0.1 and 50 cm. The slabs are thought to be backed by aluminium. It is seen that for gas layers thicker than ~ 1 cm, the cooling power at ambient temperature is larger for NH_3 than for the hydrocarbon gases. However, η^H is larger for C_2H_4 and C_2H_4O than for NH_3, implying that the hydrocarbons display a higher degree of spectral selectivity.

As already mentioned, it is of particular interest to regard gas mixtures. Figure 11 shows $\bar{e}_{s2}^H$ and η^H for $C_2H_4 + C_2H_4O$ combinations with three layer thicknesses. It is apparent that the mixtures can have a higher cooling power than either of the constituent gases, which is an important result for practical applications. The parameter η^H, on the other hand, is not strongly influenced by the mixing. The superior cooling power of the mixtures is readily understood from the spectral transmittance data, which show that a more complete coverage of the 8-13 μm interval is possible with a combination of gases than by any one of the pure gases alone. For completeness we mention that gas mixtures can also be used to provide rather high absorption across the full thermal radiation spectrum.[29]

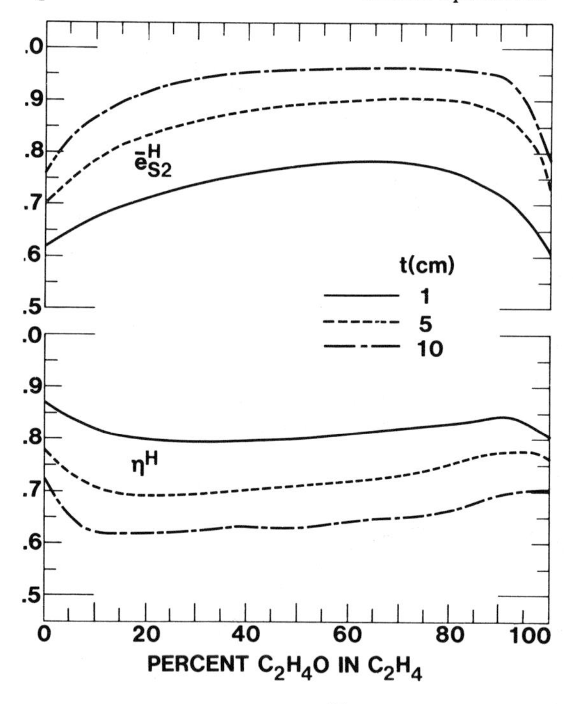

Fig. 11. Basic cooling parameters, $\bar{e}_{s2}^H$ and η^H, for mixtures of C_2H_4 and C_2H_4O having three layer thicknesses, t. (From Ref. 27).

D. Ceramic Oxide Layers

The earlier discussed approaches to selective infrared emission rely on highly reflecting surfaces covered with a solid or gaseous substance which decreases the reflectance predominantly in the 8-13 μm interval. Another possibility is to use a material which yields a high reflectance at $\lambda > 13$ μm by the Reststrahlen effect. Certain ceramic layers, approximately 1 mm in thickness, can have the desired optical properties.[4,30,31] Some further improvement may be accomplished by backing the ceramic layer with a reflecting surface.

Figure 12, based on Ref. 30, shows specular reflectance in the thermal infrared for a 1.1 mm thick layer of MgO ceramic. The material is of high density and polished to a good surface finish. The high reflectance at $13 < \lambda < 25$ μm is striking. This is the range within which most of the atmospheric radiance is impinging (cf. Fig. 1). The Reststrahlen band agrees with computations based on optical data for single crystals of MgO. Calculations predict[31] that ~ 0.5 mm thick LiF layers should have radiative properties similar to those for MgO.

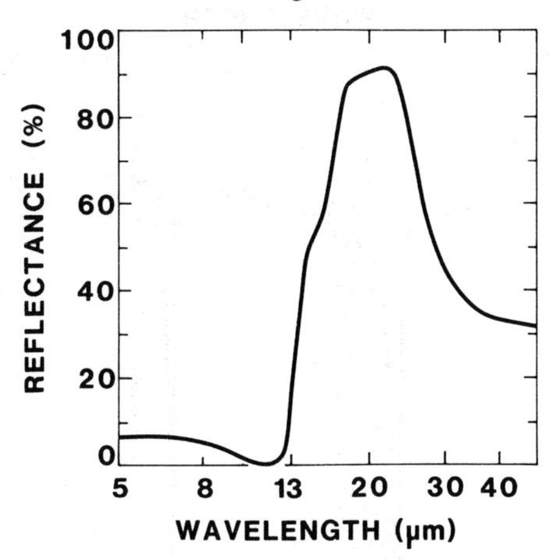

Fig. 12. Near-normal spectral reflectance for a 1.1 mm thick ceramic MgO layer backed by a reflecting surface. (Redrawn from Ref. 31).

IV. INFRARED-TRANSPARENT CONVECTION SHIELDS

Radiative cooling is governed by the difference in outgoing and incoming radiative power, as well as by the non-radiative heat influx, as discussed in Sec. II A. The non-radiative component can be specified in terms of a linear heat-transfer coefficient κ, which for free exposure to the atmosphere can be written[32]

$$\kappa = 5.7 + 3.8 \, v \tag{17}$$

in units of $Wm^{-2}K^{-1}$, where v is the wind velocity in ms^{-1}. A large temperature difference, or a large cooling power at a moderate temperature difference, requires that κ be diminished to a value on the order of 1 $Wm^{-2}K^{-1}$, which can be accomplished only by placing the infrared-emitting surface below a convection shield which is transparent in the 8-13 μm range.

A practical infrared-transparent convection shield can be constructed from high-density polyethylene foil. Figure 13 illustrates the normal transmittance through 30 μm of this material in the 5-50 μm interval. Similar data were shown in Fig. 9 above. The transmittance is ~ 85 % in the 8-13 μm range. Narrow minima at 6.8 and 13.8 μm - i.e., outside the "atmospheric window" - signify molecular absorption in the material. Oscillations at the far-infrared end of the spectrum are caused by optical interference between the parallel surfaces of the foil. Convection shields of polyethylene have been used in earlier field tests of radiative cooling.[4,5,12,25-27,31,33-35] In principle, a plate of an infrared-transparent material such as antireflection-coated Si or Ge could be employed as convection shield, but the cost would be prohibitive.

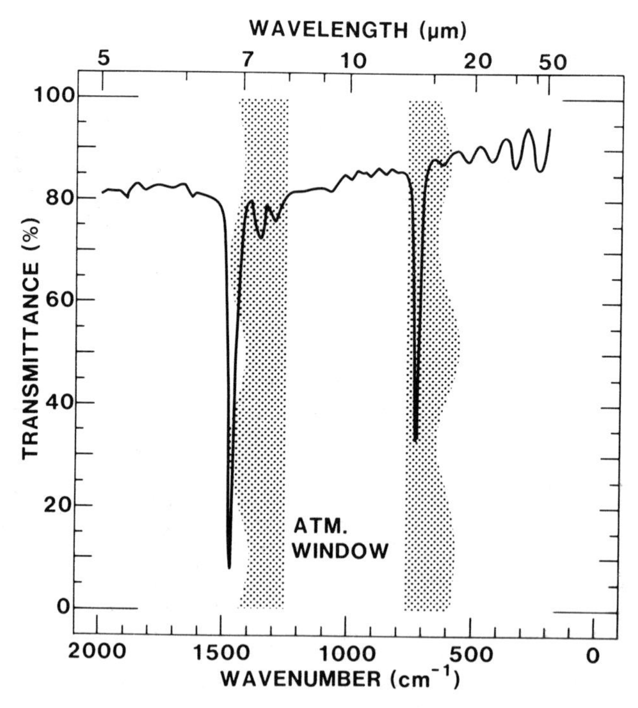

Fig. 13. Normal spectral transmittance of a 30 μm thick foil of high-density polyethylene. The 8-13 μm "atmospheric window" is indicated. (From Ref. 36).

High-density polyethylene foil has several attractive features which makes it useful for radiative cooling applications. However, there are also problems which call for materials development. First, a flexible foil is easily moved even by weak winds so that forced convection takes place, which leads to an undesired heat transfer to the radiatively cooled surface. The situation can be improved by use of a polyethylene-based cellular material, as discussed in Sec. IV A below. Second, if radiative cooling is to be utilized during the day one should try to limit the inflow of solar energy to the radiating surface. One way of doing this is by decreasing the solar transmittance of the foil, as we return to in Sec. IV B. Finally, it would be useful to have access to a polyethylene foil or plate with a lower residual absorptance in the 8-13 μm band than for the standard qualities.

A. Polyethylene-Based Cellular Material

High density polyethylene with infrared transmittance according to Fig. 13, was used for the construction of mechanically rather rigid materials comprising several layers of V-corrugated foils.[36] Figure 14 shows a sketch of a typical sample: the corrugated layers, with height h and apex angle θ, are oriented at right angles. The samples studied below have h equal to 0.5, 1 or 1.5 cm, and θ equal to 90° or 45°. The individual layers were made by hot-forming.

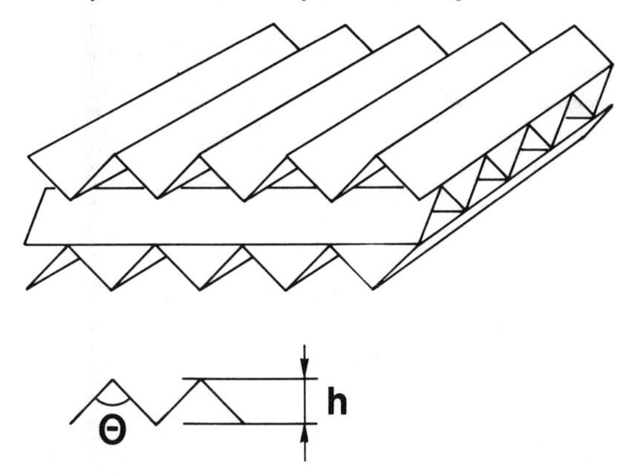

Fig. 14. The main figure shows a material with high infrared transmittance and low non-radiative heat exchange. It consists of three crossed corrugated foils of polyethylene. The lower figure indicates the pertinent parameters: the height h and the apex angle θ. (From Ref. 36).

The infrared transmittance through the material was determined by use of the experimental arrangement shown in the inset of Fig. 15. The sample was placed closely in front of a blackbody-like radiator and was viewed by infrared-imaging equipment. This type of measurement made it possible to evaluate the transmittance as a function of the number of individual foils. Filled circles and open triangles in Fig. 15 pertain to corrugated foils with different apex angles. The transmittance goes monotonically from ~ 90 % for a single foil to ~ 60 % for five foils, which is the expected behaviour. Data for flat foils, indicated by open squares, were in acceptable agreement with spectrophotometric recordings. It is found from Fig. 15 that the transmittance is higher for corrugated foils than for flat ones; this is corroborated by recent literature data.[37]

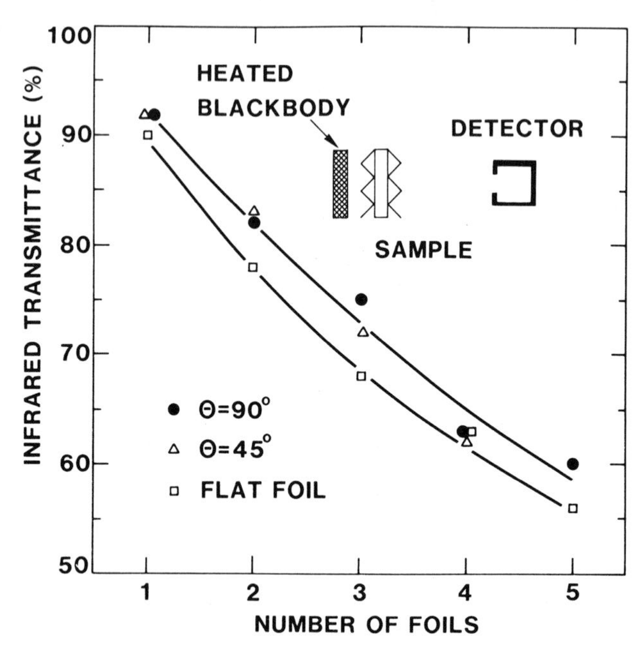

Fig. 15. Transmittance in the 8-13 μm range vs. number of polyethylene foils for different sample geometries. (From Ref. 36).

The non-radiative thermal resistance was measured by a modifiction of the well known guarded hot-plate technique. The inset of Fig. 16 shows the essential features of this technique. Highly reflecting aluminium foil was attached to the plates in order to minimize direct radiative coupling between them. Figure 16 shows non-radiative thermal resistance as a function of sample thickness. The data for corrugated foils yield a consistent pattern with a monotonically varying thermal resistance lying below that of the values for empty air gaps. Heat flow through partially transparent materials is complicated, and the shown data are not readily amenable to theoretical modelling.

Looking at Figs. 15 and 16, it is evident that an increase of the thermal resistance is accompanied by a decrease of the infrared transmittance. With regard to radiative cooling applications, there is thus a trade-off between the two properties. As a characteristic result it is found that at a thickness of 4.5 cm - corresponding to three 1.5 cm-thick corrugated foils - the thermal resistance is 1.1 m^2KW^{-1} and the infrared transmittance is 73 %. Further information on heat transfer through cellular materials is given in the chapter on Transparent Insulation Materials.

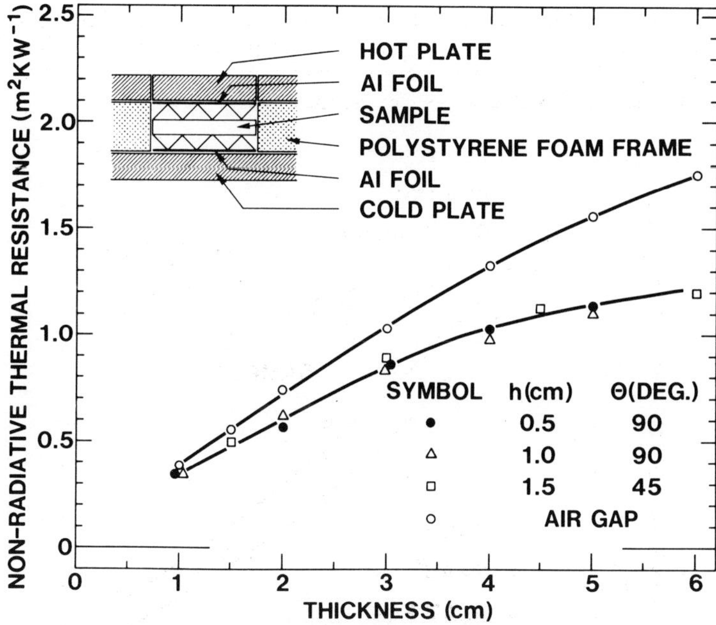

Fig. 16. Non-radiative thermal resistance versus thickness for different sample geometries. (From Ref. 36).

B. Polyethylene-Based Solar Reflecting Foils

It is possible to limit the solar transmittance of polyethylene foils, while their transmittance in the 8-13 μm range is conserved, by applying a coating or by pigmentation of the material. With regard to coatings, it is possible to use a semiconductor with a cut-off wavelength larger than the infrared end of the solar spectrum (~ 2.5 μm). Among the interesting materials are tellurium and PbTe (cf. Refs. 26 and 38). If the thickness is > 0.05 μm, the solar transmittance will be low (< 0.01 at λ = 0.5 μm). The refractive indices of these coatings are high, though, which tends to limit the transmittance in the 8-13 μm band. Three design possibilities exist: The first of these is to use a very thin layer. Curve 1 in Fig. 17, referring to 0.05 μm of tellurium, shows that the transmittance is ~ 87 % within the "atmospheric window". The second possibility is to have a thickness such that the film serves as a $\lambda/2$-layer for $\lambda \approx 11$ μm. A tellurium film meeting this condition should be 1.1 μm thick; its calculated transmittance is shown by curve 2 in Fig. 17. It is found that the transmittance is high only in part of the "atmospheric window". The third and principally best design is to embed the tellurium film between two layers which antireflect in the 8-13 μm range. Curve 3 in Fig. 17 pertains to the calculated transmittance for 0.7 μm of tellurium between 0.88 μm of CdTe and 1.0 μm of ZnS. The transmittance is seen to be high within the whole "atmospheric window".

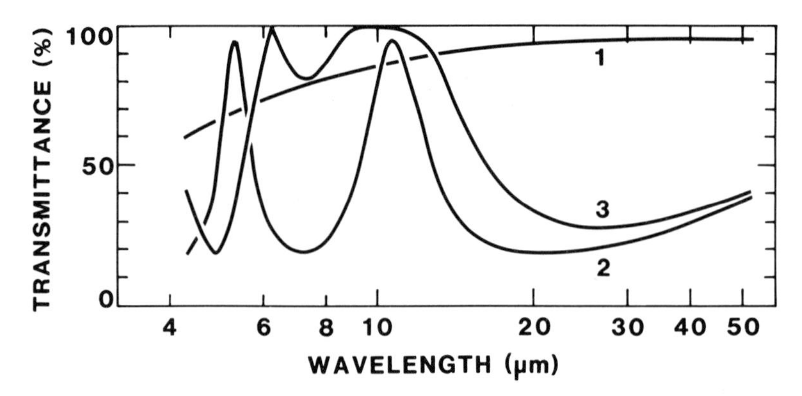

Fig. 17. Calculated infrared transmittance through tellurium based films. The curves 1, 2 and 3 refer to 0.05 μm tellurium, 1.1 μm tellurium, and 0.88 μm CdTe/0.7 μm Te/1.0 μm ZnS, respectively. The effect of an underlying polyethylene substrate is not included in the computation. (Redrawn from Ref. 26).

Pigmentation or dyeing of the polyethylene foils offers another possibility to combine solar rejection with thermal infrared transmittance.[39-42] Figures 18 and 19 show some measured[42] reflectance and transmittance data for 100-μm-thick polyethylene foils pigmented with 0.23-μm-diameter TiO_2 (rutile) particles to the shown volume fractions (f). It is seen that a rather high reflectance of solar energy (cf. Fig. 18) can be combined with some transmittance in the 8-13 μm range (cf. Fig. 19). The properties are not ideal for radiative cooling applications, though, and a pigment whose phonon absorption lies at longer wavelengths than that for TiO_2 would be better. The optical properties of pigmented foils can be modeled,[42,43] at least semiquantitatively, by the theory for radiation scattering that was introduced in the chapter on Optical Properties of Inhomogeneous Two-Component Materials. In order to diminish the solar transmittance it may be favourable to combine a white TiO_2-containing foil with a downwards-facing solar absorbing foil, which can consist of polyethylene containing carbon black. Figure 20 shows spectral transmittance for such a tandem foil.[41] It appears that a solar transmittance of ~ 0.1 can be combined with a transmittance of ~ 0.7 for the 8-13 μm range.

V. RESULTS OF SOME FIELD TESTS

Several field tests have been conducted with radiatively cooled devices. They have shown unambiguously that radiative exchange with the clear sky is a viable means for reaching low temperatures. It is fair to say, though, that only a few of the devices have taken full advantage of the cooling potential. A series of experiments have been performed to elucidate the cooling power and attainable temperature difference using selectively infrared-emitting surfaces mounted under simple transparent convection shields of polyethylene. Some examples of test results are reported in Sec. V A. In order to reach very low temperatures, one

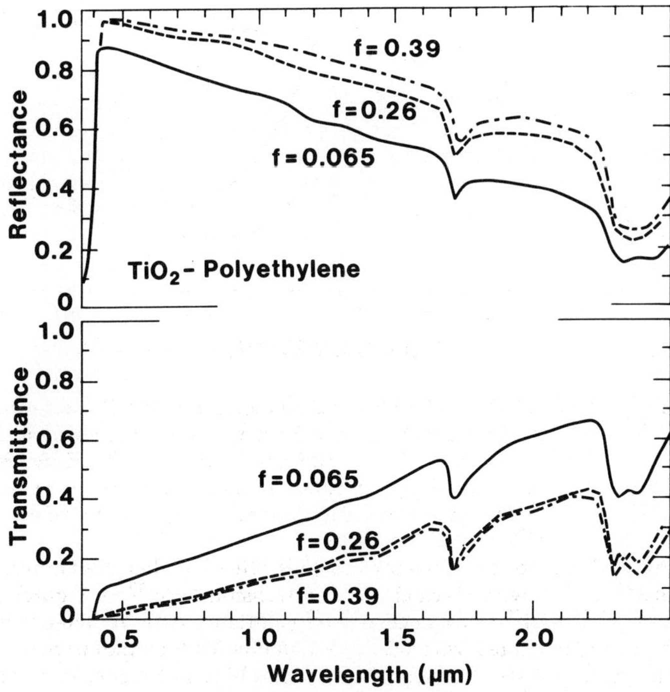

Fig. 18. Total reflectance and total transmittance for TiO_2-polyethylene foils with pigment volume fraction, f, as given in the figure. (From Ref. 42).

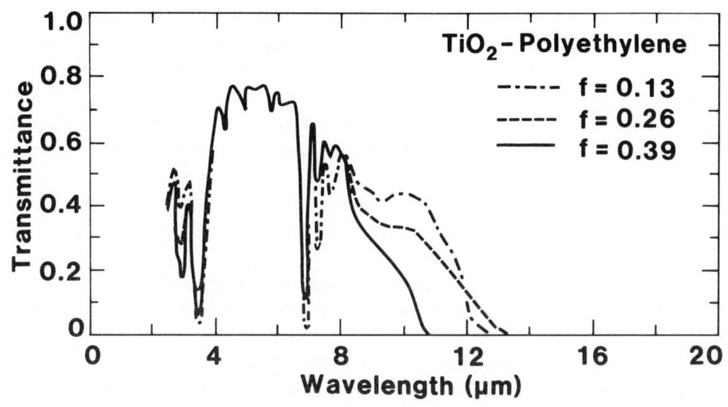

Fig. 19. Total transmittance in the thermal-infrared wavelength region for TiO_2-polyethylene foils with pigment volume fraction, f, as given in the figure. (From Ref. 42).

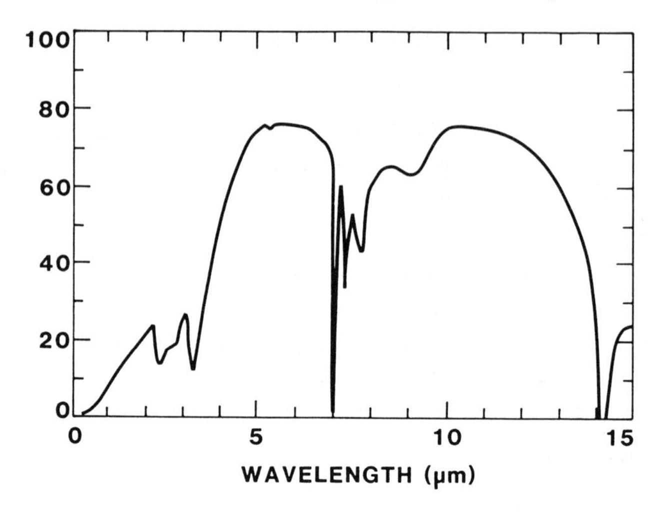

Fig. 20. Transmittance of a double-face black-and-white foil of pigmented polyethylene. (Redrawn from Ref. 41).

can work with multistage cooling devices, as illustrated in Sec. V B. Solar reflecting convection shields offer a very simple approach to radiative cooling. Some initial results based on this idea are given in Sec. V C.

A. Selectively Infrared-Emitting Surfaces Placed under Transparent Convection Shields: Some Examples

A number of selectively infrared-emitting surfaces were discussed in Sec. III, and silicon-oxynitride coatings, backed by highly reflecting aluminium, were found to have very good spectral properties with low reflectance in the 8-13 μm band and high reflectance for other wavelengths. Such surfaces have been integrated in cooling devices[44] whose general design is illustrated in Fig. 21. The cooling plate, 0.25 m^2 in size, is of aluminium and has a 1.2 μm thick coating of $SiO_{1.47}N_{0.54}$ produced by reactive radio frequency magnetron sputtering. The near-normal spectral reflectance of the surface was shown in Fig. 4c. The cooling plate was mounted under a single high-density polyethylene foil with transmittance according to Fig. 13. An arrangement with stretching ribs held the foil in tension. Thermal emission from the walls surrounding the cooling plate was minimized by the use of aluminized Mylar foil. Thermal insulation of the cooling plate was accomplished by polyurethane foam. The underside of the cooling plate was equipped with thermometers and electrical heating elements.

Cooling tests were conducted under a variety of climatic conditions. Figure 22 shows the temperature which was recorded for free exposure to the clear sky during a winter night in Gothenburg, Sweden. The dew-point temperature is

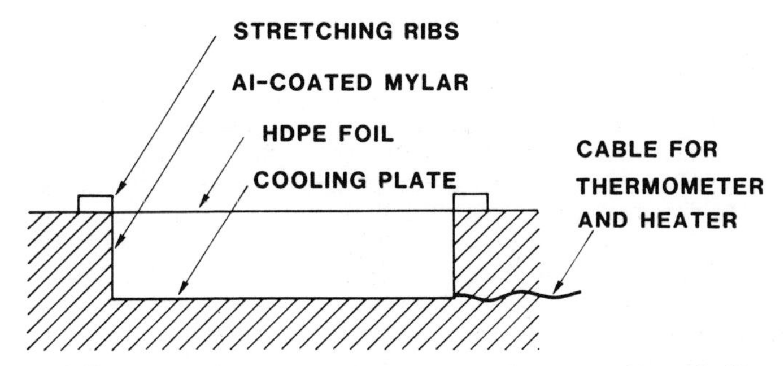

Fig. 21. Sketch of a device for testing radiative cooling of silicon-oxy-nitride coated surfaces.

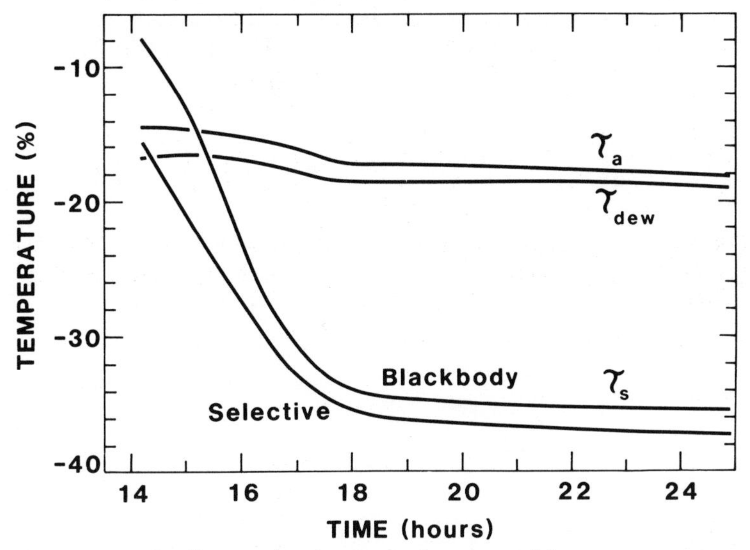

Fig. 22. Cooling performance vs. time for ambient temperature τ_a, dewpoint temperature τ_{dew}, and for the surface temperature τ_s of materials with an infrared selective silicon-oxy-nitride coating and with a blackbody-like paint layer. The data were taken with devices of the kind illustrated in Fig. 21.

~ - 18 ° C, and hence the test simulates to some extent the cooling performance pertinent to a very dry climate. The curves denoted "selective" and "blackbody" indicate the temperature vs. time for two cooling devices incorporating an infrared-selective silicon-oxynitride surface and, for reference, a blackbody-like painted surface. It is seen that the infrared-selective surface reached $\Delta T \approx 20°C$ whereas the blackbody-like surface reached $\Delta T \approx 18.5°C$. The observed temperature difference, and its dependence on the net cooling power, is consistent with calculations[44] based on the LOWTRAN computer code and a model atmosphere of the Subarctic Winter type. Much larger temperature drops, and larger differences between the selective and non-selective surfaces, are expected in devices with lower non-radiative heat influx.

Gas slabs backed by metal offer another possibility to exploit radiative cooling, as discussed in Sec. III C. Practical testing was performed[27] with cooling panels of the type shown in Fig. 23. They comprise a polystyrene box with an infrared-transparent window consisting of three polyethylene foils. The interior of the box was clad with aluminium foil. Depressurized gas at ambient temperature was run via a flowmeter through a 10 cm thick region under the infrared-transparent window. The area exposed to the sky was 0.38 m^2. Thermometers were arranged to measure the gas temperature at the inlet and outlet.

Figure 24 shows an example of results from a cooling test with C_2H_4 gas. The experiment was carried out in full daylight with a small screen set to block the direct solar radiation. The ensuing load due to diffuse radiation is expected to be on the order of 10 Wm^{-2}. The air temperature was ~ 5°C and the relative humidity was ~ 40 %. At point A in the figure the gas was introduced at a rate of 0.6×10^{-3} m^3 s^{-1}. A temperature difference of 7°C was established within a few minutes. At point B the gas flow was stopped at which point ΔT became as large as 10°C. Resumed gas flow at point C again led to $\Delta T \approx 7$ °C. Devices using self-circulation of radiatively cooled gas have been tested.[45]

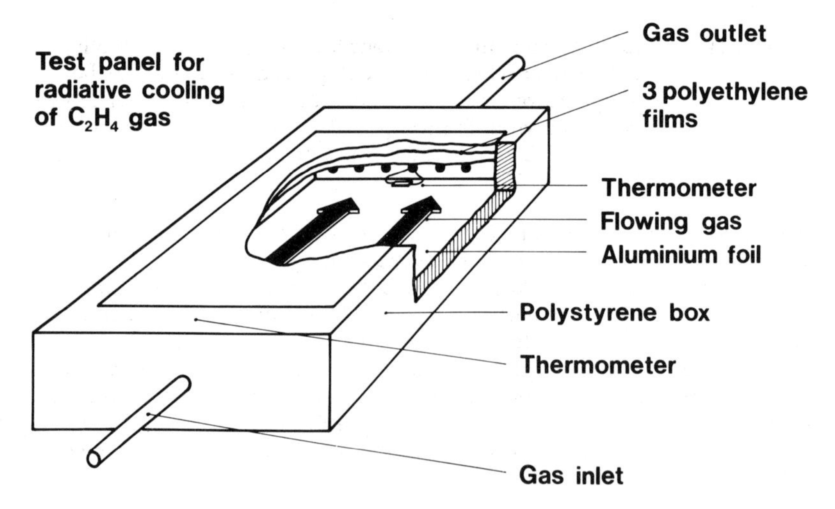

Fig. 23. Cutaway diagram of panel for testing radiative gas cooling. The large arrows indicate laminar gas flow. (From Ref. 27).

Certain ceramics can display infrared selectivity as a consequence of their Reststrahlen band. MgO is such a material, as pointed out in Sec. III D. Test results for 1.1 mm thick polished MgO ceramic plates, ~ 0.5 m^2 in size, have been reported.[31] The plates were well insulated on the underside and mounted below

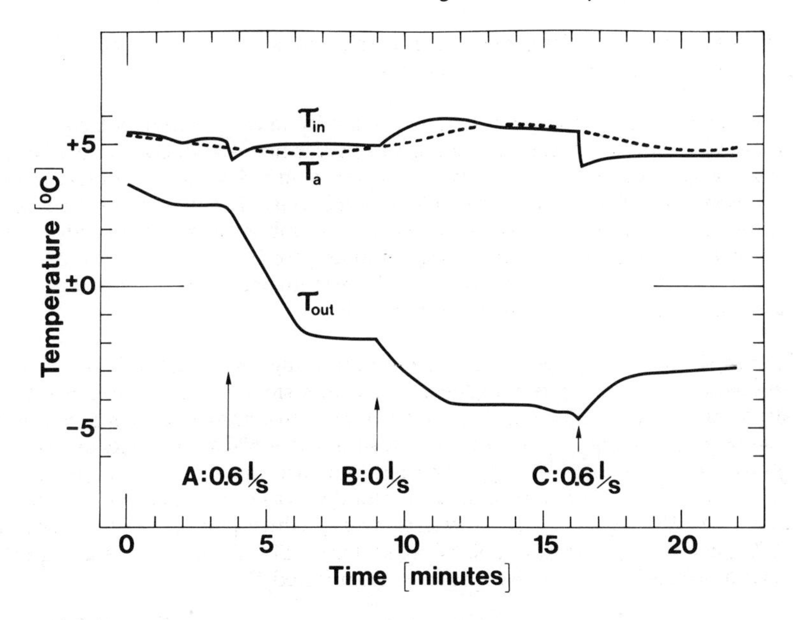

Fig. 24. Excerpt from an experiment for testing radiative cooling with a 10-cm-thick slab of C_2H_4. τ_{in} and τ_{out} denote the temperatures at the gas inlet and outlet, respectively, in the panel shown in Fig. 23. τ_a is the ambient temperature. At A, B, and C, the gas flow was adjusted to the shown rates. (From Ref. 27).

a 50 μm thick polyethylene foil. They could be heated electrically. Figure 25 shows measured temperature vs. time for cooling devices incorporating a MgO plate and, for reference, a blackbody-like painted surface. A temperature more than 20°C lower than τ_a was measured on the MgO plate. The blackbody-like surface could reach a temperature difference which was ~ 3°C smaller. The difference between the two types of surfaces increased to ~ 5°C when the distance between the cooling surface and the convection shield was increased from the original 2.5 cm up to 5 cm. The difference in performance between the two types of surfaces became smaller as power was fed into the heater, which is the expected result.

B. Multistage Cooling Devices

Carefully constructed multistage devices can be used to create very low temperatures on small surfaces exposed to atmospheres with low humidity. A number of important experiments with such devices were conducted at high altitude in the French Pyrennées[4,46] and in the Atacama desert in Chile[47] during the 1960's and 1970's.

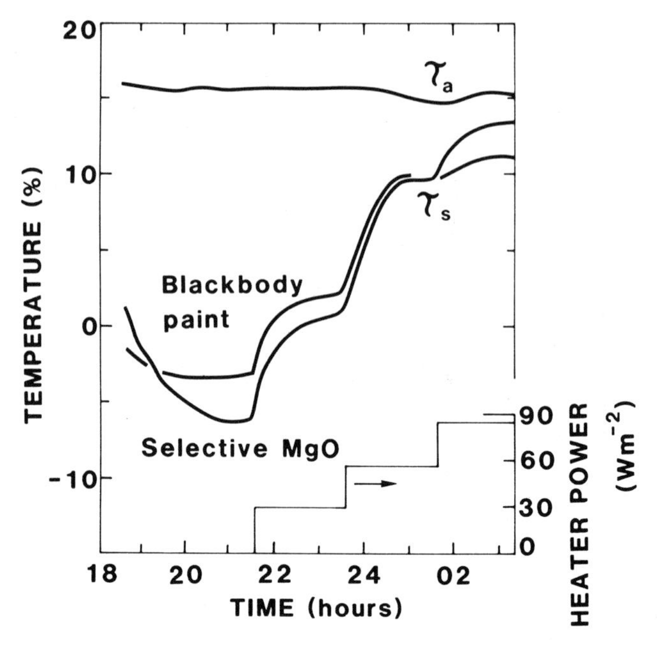

Fig. 25. Cooling performance vs. time for ambient temperature τ_a and for the surface temperature τ_s of infrared-selective MgO ceramic and of black-body-like paint. The result of an input of electric power to the surfaces is shown. (Redrawn from Ref. 31).

Figure 26 illustrates a five-stage device[46] with 1.2 m base side. Numbers 1-5 denote infrared-radiating surfaces of oxidized aluminium or TiO_2 paint. Numbers 6-10 denote 30 μm thick polyethylene foils. The innermost cooling surface, having an area of 25 cm^2, could reach $\Delta T \approx 37°C$ under favourable meteorological conditions despite the fact that the radiating surfaces must have been rather blackbody-like. If the interior of the cooling device was filled with CO_2 gas, a temperature difference exceeding 40°C was observed.

Figure 27 illustrates the thermal performance of a four-stage device analogous to the one shown in Fig. 26. The different curves denote τ_a and the temperatures τ_s of the cooling surfaces. The relative humidity was about 50 %. An arrow marks the introduction of CO_2 gas. A maximum temperature difference of ~ 33°C was observed in this experiment.

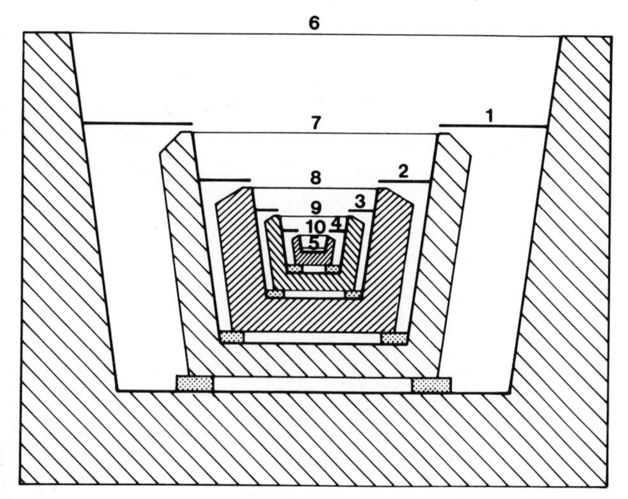

Fig. 26. Sketch of a five-stage cooling device. (From Ref. 46).

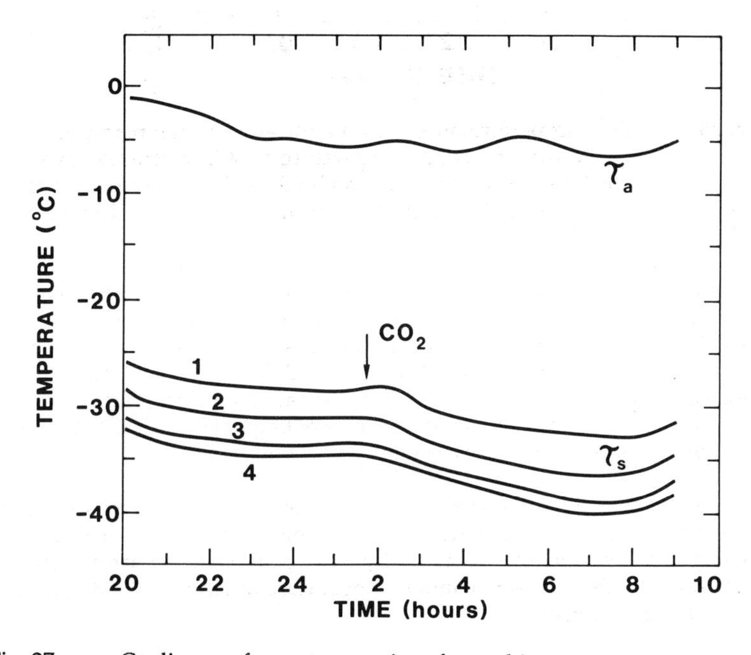

Fig. 27. Cooling performance vs. time for ambient temperature τ_a and for the temperature τ_s of four successively smaller and colder surfaces arranged in a four-stage cooling device similar to the one in Fig. 26. And arrow marks the introduction of CO_2 gas. (Redrawn from Ref. 46).

C. Cooling with Polyethylene-Based Solar Reflecting Foils

Cooling can be accomplished even during the day by placing a foil with high solar reflectance and significant transmittance in the 8-13 μm range over the space to be cooled. Such foil materials have been discussed in the literature, and the results of some development work were mentioned in Sec. IV B. The foils were used[39-41,48] for space cooling. Figure 28 shows the temperature inside and outside a greenhouse constructed with foils according to Fig. 20. A temperature difference of 3-4°C, with the interior being the colder, is observed both in the day and the night. Hence there is no doubt as to the feasibility of space cooling by use of solar reflecting and infrared transmitting foils. The studied foil material does not exhibit optimized properties, though, and it is expected that a significant improvement could be accomplished if the foils had a higher solar reflectance.

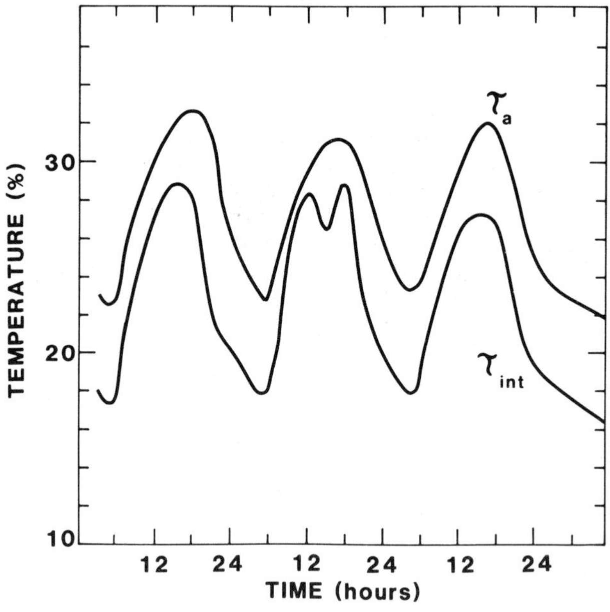

Fig. 28. Cooling performance vs. time for ambient temperature τ_a and for the temperature τ_{int} inside a greenhouse constructed with solar reflecting and infrared transmitting foil. (Redrawn from Ref. 48).

VI. CONCLUDING REMARKS

Radiative cooling is an ubiquitous phenomenon under a clear sky. The power available for cooling is ~ 100 Wm^{-2} at ambient temperature. To put this number in perspective, we point out that the maximum solar energy available for heating is ~ 1000 Wm^{-2}, i.e., only one order of magnitude larger.

The relation between cooling power and temperature difference can be understood quantitatively if the following three properties are specified:

(i) The atmospheric downwards radiance should be known. Often one can use data for model atmospheres to simulate results for cloud-free skies. An alternative is to rely on radiosoundings, which are done on a regular basis at most airports.

(ii) The spectral emittance of the radiating surface should be known. Spectrophotometric data for the reflectance at a 45° angle of incidence yield useful information.

(iii) The non-radiative heat influx to the cooling surface should be known. Practical experience tells that this limiting factor is difficult to come to grips with, and it appears that most practical devices are characterized by larger heat transfer coefficients than predicted from simple theories of thermal insulation involving stationary non-convecting gases.

An infrared-selective surface with large emittance in the 8-13 μm range and low absorptance elsewhere is capable of yielding the lowest temperature in a well-insulated device. Such surfaces have been produced and optimized as discussed in this chapter. At ambient temperature the highest cooling power is obtained with a blackbody-like surface, and whether an infrared-selective surface offers any advantage in a device can only be judged when the required cooling performance and the non-radiative heat influx are specified. In many practical cases it turns out that a cheap blackbody-like surface is, in fact, the best option. On the other hand it should be remembered that an infrared-selective gas has advantages with regard to transport and heat exchange of the coolant. This gas must be safely encased in a tight but infrared transparent container, which may be difficult to achieve under practical conditions.

A better infrared-transparent convection shield is needed if radiative cooling to low temperatures is going to be generally applicable. It appears that polyethylene is the only viable alternative and efforts should be focussed on trying to make more mechanically stable materials without sacrificing too much infrared transmittance. Foils with a supporting metal grid as well as extremely high-density foils with minimized absorption in the 8-13 μm band are of interest. Another important issue is to produce pigmented foils with a higher solar reflectance than the one measured for previously produced materials.

In a global perspective the need for cooling is as important as the need for heating. Radiative cooling has almost innumerable potential applications. Cold storage of food is of obvious importance and can be accomplished with simple radiatively cooled devices. Deep freezing, on the other hand, is feasible only under special climatic conditions. Climatization of buildings,[33,34,38-41,48] desalination of water by freezing,[2] and condensation of atmospheric humidity are other important applications. In fact, condensation irrigation has been tested[49,50] with radiatively cooled surfaces, and about 0.4 litres of water has been extracted per m^2 of radiating surface and per night in different climates. The list of possible applications can be continued.

As a final point we remark that it is surprising that the development of materials and devices for exploiting radiative cooling is still in its infancy. It is our opinion that radiative cooling offers many interesting and important opportunities for fundamental and applied research.

REFERENCES

1. F. Arago, Annuaire du Bureau des Longitudes pour l'An 1828, p. 149; reprinted in *Oeuvres Completes de François Arago*, Vol. 8 (Gide & Weigel, Paris and Leipzig, 1858), p. 87.
2. M.N. Bahadori, Sci. Am. 238, 144 (1978); in *Solar Energy Conversion: An Introductory Course*, edited by A.E. Dixon and J.D. Leslie (Pergamon, New York, 1979), p. 461.
3. A.K. Head, Australian Patent No. 239364 (1959); U.S. Patent No. 3043112 (1962).
4. F. Trombe, Rev. Gen. Therm. 6, 1285 (1967).
5. C.G. Granqvist and A. Hjortsberg, J. Appl. Phys. 52, 4205 (1981).
6. M. Martin, in *Passive Cooling*, edited by. J. Cook (MIT Press, Cambridge, 1989), p. 138.
7. K. Ya. Kondratyev, *Radiation in the Atmosphere* (Academic, New York, 1969).
8. V.E. Zuev, *Propagation of Visible and Infrared Radiation in the Atmosphere* (Wiley, New York, 1974).
9. G. Herzberg, *Molecular Spectra and Molecular Structure II. Infrared and Raman Spectra of Polyatomic Molecules* (Van Nostrand, Princeton, 1945).
10. E.E. Bell, L. Eisner, J. Young and R.A. Oetjen, J. Opt. Soc. Am. 50, 1313 (1960).
11. P. Berdahl and R. Fromberg, Solar Energy 29, 299 (1982).
12. S. Brunold, Untersuchungen zum Potential der Strahlungskühlung in Ariden Klimatzonen, Thesis, Fraunhofer Institut für Solare Energiesysteme, Freiburg, Germany, 1989 (unpublished).
13. F.X. Kneizys, E.P. Shettle, W.O. Gallery, J.H. Chetwynd, Jr., L.W. Abreu, J.E.A. Selby, R.W. Fenn, and R.A. McClatchey, Atmospheric Transmittance/ Radiance Computer Code LOWTRAN 5, Air Force Geophysics Laboratory, Technical Report AFGL-TR-80-0067 (February 1980).
14. T.S. Eriksson and C.G. Granqvist, Appl. Opt. 21, 4381 (1982).
15. M. Martin and P. Berdahl, Solar Energy 33, 321 (1984).
16. S.B. Idso, Water Resources Res. 17, 295 (1981).
17. I. Hamberg, J.S.E.M. Svensson, T.S. Eriksson, C.G. Granqvist, P. Arrenius and F. Norin, Appl. Opt. 26, 2131 (1987).
18. C.G. Ribbing, Opt. Lett. 15, 882 (1990).
19. T.S. Eriksson and C.G. Granqvist, Appl. Opt. 22, 3204 (1983).
20. T.S. Eriksson and C.G. Granqvist, J. Appl. Phys. 60, 2081 (1986).
21. D.E. Aspnes and J.B. Theeten, J. Appl. Phys. 50, 4928 (1979).
22. T.S. Eriksson, S.-J. Jiang and C.G. Granqvist, Solar Energy Mater. 12, 319 (1985).
23. M.D. Diatezua, A. Dereux, A. Ronda, J.P. Vigneron, P. Lambin and R. Caudano, Proc. Soc. Photo-Opt. Instrum. Engr. 1149, 80 (1989).
24. T.S. Eriksson, A. Hjortsberg and C.G. Granqvist, Solar Energy Mater. 6, 191 (1982).

25. S. Catalanotti, V. Cuomo, G. Piro, D. Ruggi, V. Silvestrini and G. Troise, Solar Energy 17, 83 (1975).
26. Ph. Grenier, Rev. Phys. Appl. 14, 87 (1979).
27. E.M. Lushiku and C.G. Granqvist, Appl. Opt. 23, 1835 (1984).
28. L.J. Bellamy, *The Infra-Red Spectra of Complex Molecules* (Chapman & Hall, London, 1975).
29. T.S. Eriksson, C.G. Granqvist and J. Karlsson, Solar Energy Mater. 16, 243 (1987).
30. B. Piriou, Rev. Int. Hautes Temp. Refract. 10, 283 (1973).
31. P. Berdahl, Appl. Opt. 23, 370 (1984).
32. J.A. Duffie and W.A. Beckman, *Solar Energy Thermal Processes* (Wiley, New York, 1974), p.83.
33. D. Michell and K.L. Biggs, Appl. Energy 5, 263 (1979).
34. B. Landro and P.G. McCormick, Int. J. Heat Mass Transfer 23, 613 (1980).
35. P. Berdahl, M. Martin and F. Sakkal, Int. J. Heat Mass Transfer 26, 871 (1983).
36. N.A. Nilsson, T.S. Eriksson and C.G. Granqvist, Solar Energy Mater. 12, 327 (1985).
37. T. Ishibashi and M. Ishida, in *Solar World Forum*, edited by D.O. Hall and J. Morton (Pergamon, Oxford, 1982) Vol. 1, p. 198.
38. S. El Golli and Ph. Grenier, J. Phys. (Paris) 42, C1-431 (1981).
39. A. Addeo, E. Monza, M. Peraldo, B. Bartoli, B. Coluzzi, V. Silvestrini and G. Troise, Nuovo Cimento 1C, 419 (1978).
40. A. Addeo, L. Nicolais, G. Romeo, B. Bartoli, B. Coluzzi and V. Silvestrini, Solar Energy 24, 93 (1980).
41. A. Andretta, B. Bartoli, B. Coluzzi and V. Cuomo, J. Phys. (Paris) 42, C1-423 (1981).
42. G.A. Niklasson and T.S. Eriksson, Proc. Soc. Photo-Opt. Instrum. Engr. 1016, 89 (1988).
43. T.M.J. Nilsson and G.A. Niklasson, Gothenburg Institute of Physics Report GIPR-299, 1991 (unpublished).
44. T.S. Eriksson, Ph. D. Thesis, Dept. of Physics, Chalmers University of Technology, Gothenburg, Sweden, 1985 (unpublished).
45. E.M. Lushiku, Ph. D. Thesis, Dept. of Physics, University of Dar es Salaam, Dar es Salaam, Tanzania, 1986 (unpublished).
46. F. Trombe, A. Le Phat Vinh and M. Le Phat Vinh, J. Rech. CNRS 65, 563 (1964).
47. G.S. Castellanos and J. Fournier, Chauffage, Ventilation, Conditionnement 3, 31 (1975).
48. B. Bartoli and V. Silvestrini, Le Scienze 139, 70 (1980).
49. J. Rademaekers, T.S. Eriksson and E.M. Lushiku, unpublished.
50. E.M. Lushiku and R.T. Kivaisi, Proc. Soc. Photo-Opt. Instrum. Engr. 1149, 111 (1989).